Get the most from this book

Welcome to your Revision Guide for the WJEC GCSE Mathematics Foundation courses. This book will provide you with sound summaries of the knowledge and skills you will be expected to demonstrate in the exams, with additional hints and techniques on every page. Throughout the book, you will also find a wealth of additional support to ensure that you feel confident and fully prepared for your GCSE Maths Foundation examinations.

This Revision Guide is divided into four main sections, with additional support at the back of the book. The four main sections cover the four mathematical themes that will be covered in your courses and examined: Number, Algebra, Geometry & Measures and Statistics & Probability.

Features to help you succeed

Each theme is broken down into one-page topics as shown in this example:

The knowledge you have learned on your course is reduced to the key rules for this topic area. You will need to understand and remember these for your exam.

Worked examples are provided to remind you how the rules work. Each rule is highlighted next to where it is being used.

Exam-style questions provide real practice on the topic area, with allocated marks so you can see the level of response that is required.

Each page is given a level of difficulty so you can understand the level of challenge. Broadly speaking, Low denotes grades 1-2, Medium = 3, High = 4-5.

Areas where common errors are often made are highlighted to help you avoid making similar mistakes.

These are the terms and phrases you will need to remember for this topic.

Hints and techniques will suggest what to remember or how to approach an exam question.

Setting up and solving simple equations

REVISED

MEDIUM

Algebra

Rules

1. Always use the inverse operations to solve an equation.
2. + and − are the inverse of one another.
3. × and ÷ are the inverse of one another.
4. To set up an equation a variable must be defined.

Worked examples

a 2 $2p + 5 = 17$

   $2p + 5 − 5 = 17 − 5$ [subtract 5 from each side of the equation]

   [−5 is the inverse operation of + 5]

   3 $2p = 12$ [÷ 2 is the inverse of × 2]

   $p = 6$ [divide each side of the equation by 2]

b 2 Ann is two years younger than Ben.
   Clara is twice as old as Ben.
   The total of their ages is 58.
   Work out their ages.

   Answer
   Let Ben's age be x

   Ann's age will be $x − 2$

   Clara's age will be $2x$

   Firstly, set up the equation:

   $x + (x − 2) + 2x = 58$

   Now collect like terms:

   $4x − 2 = 62$

   $4x = 60$ [Add 2 to each side]

   $x = 15$ [Divide each side by 4]

   Ann is 13, Ben 15 and Clara 30.

Key terms
Equation
Inverse operation
Solve
Variable

Exam tips
Always use algebraic methods and show your working to gain full marks.
Always check your answer to make sure it is correct.

Exam-style questions

1 Solve these equations.
   a $a − 3 = 7$ **[1]**   b $\frac{b}{5} = 3$ **[1]**   c $3c + 9 = 7$ **[2]**
   d $\frac{5d}{2} + 4 = 29$ **[2]**   e $6 − 2e = 3$ **[2]**

2 Here is a rectangle.
   The length is $2x + 5$.
   The width is $x − 3$.
   The perimeter is 46 cm.
   Work out the area of the rectangle in cm². **[4]**

   $2x + 5$
   $x − 3$

CHECKED ANSWERS

Each theme section also includes the following:

Pre-revision check

Each section begins with a test of questions covering each topic within that theme. This is a helpful place to start to see if there are any areas which you may need to pay particular attention to in your revision. To make it easier, we have included the page reference for each topic page next to the question.

Exam-style question tests

There are two sets of tests made up of practice exam-style questions to help you check your progress as you go along. These can be found midway through and at the end of each theme. You will find **Answers** to these tests at the back of the book.

At the end of the book, you will find some very useful information provided by our assessment experts:

The language used in mathematics examinations

This page explains the wording that will be used in the exam to help you understand what is being asked of you. There are also some extra hints to remind you how to best present your answers.

Exam technique and formulae that will be given

A list of helpful advice for both before and during the exams, and confirmation of the formulae that will be provided for you in the exams.

Common areas where students make mistakes

These pages will help you to understand and avoid the common misconceptions that students sitting past exams have made, ensuring that you don't lose important marks.

One week to go

Reminders and formulae for you to remember in the final days before the exam.

Tick to track your progress

Use the revision planner on pages v to vi to plan your revision, topic by topic. Tick each box when you have:
- worked through the pre-revision check
- revised the topic
- checked your answers.

You can also keep track of your revision by ticking off each topic heading in the book. You may find it helpful to add your own notes as you work through each topic.

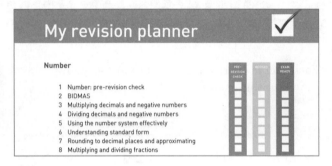

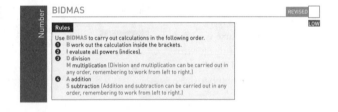

My revision planner

Geometry and Measures

Statistics and Probability

Exam preparation

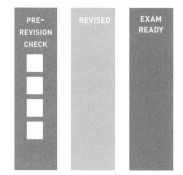

PRE-
REVISION
CHECK

REVISED

EXAM
READY

Number: pre-revision check

Check how well you know each topic by answering these questions. If you get a question wrong, go to the page number in brackets to revise that topic.

1 Work out
  a $5 + 7 \div 2$
  b 2×3^2
  c $\frac{16 - 4.1}{2.6 + 5.9}$ (page 2)

2 Work out
  a 3.174×8
  b 12.6×4.5 (page 3)

3 Work out
  a $10.44 \div 0.4$
  b $42.208 \div 1.6$ (page 4)

4 Work out
  a 1.8×0.01
  b $12.45 \div 0.1$
  c 254.9×1000
  d $0.00487 \div 0.0001$ (page 5)

5 a Write 0.00572 in standard form.
  b Write 3.184×10^4 as an ordinary number.
  (page 6)

6 Write 16.3548 correct to
  a 1 decimal place
  b 2 decimal places
  c 3 decimal places.   (page 7)

7 Write down an estimate for the value of
  a 3201
  b 0.0029   (page 7)

8 Work out
  a $\frac{4}{9} \times \frac{5}{8}$

  b $\frac{1}{3} \times \frac{4}{11} \times \frac{9}{20}$

  c $6 \div \frac{2}{3}$

  d $\frac{15}{42} \div \frac{5}{7}$ (page 8)

9 Work out
  a $\frac{7}{12} - \frac{2}{5}$

  b $3 + \frac{3}{4} - \frac{1}{3}$

  c $10\frac{2}{3} + 7\frac{3}{5}$

  d $3\frac{4}{7} \times 2\frac{4}{5}$

  e $2\frac{1}{10} \div 1\frac{3}{4}$ (page 9)

10 a Convert to a percentage
  i $\frac{11}{20}$
  ii 1.9
  b Convert 6% to a
  i fraction
  ii decimal.   (page 10)

11 a Increase £12.50 by 4%.
  b Decrease 550 m by 32%.   (page 11)

12 a Work out 20 cm as a percentage of 2.5 m.
  b A clock is bought for £72 and sold for £54. What is the percentage loss?   (page 12)

13 a Divide 280 m in the ratio 3 : 5.
  b Alfie, Bernice and Charlie share a sum of money in the ratio 1 : 3 : 5. What fraction of the money does Charlie get?   (page 14)

14 Five identical spheres weigh a total of 1.235 kg. Find the total weight of eight of these spheres.   (page 15)

15 Write each of these as a power of 3
  a $(3^4 \times 3^3) \div 3^2$
  b $3^8 \div (3^2 \times 3)$
  c $(3^5 \times 3^4)^2$ (page 16)

16 Write 1260 as a product of its prime factors. (page 17)

BIDMAS

Rules

Use **BIDMAS** to carry out calculations in the following order.
1 **B** work out the calculation inside the brackets.
2 **I** evaluate all powers (indices).
3 **D** division
   M multiplication (Division and multiplication can be carried out in any order, remembering to work from left to right.)
4 **A** addition
   S subtraction (Addition and subtraction can be carried out in any order, remembering to work from left to right.)

Worked examples

a   Work out $2^4 \times 10 + (15 - 7) \div 4$

   Answer
   1 $= 2^4 \times 10 + 8 \div 4$ $(15 - 7 = 8)$
   2 $= 16 \times 10 + 8 \div 4$ $(2^4 = 16)$
   3 $= 160 + 2$ $(16 \times 10 = 160$ and $8 \div 4 = 2)$
   4 $= 160 + 2 = 162$

b   Use your calculator to work out $\frac{5.63 + 12.17}{19.26 - 4.9}$

   Answer
   $\frac{5.63 + 12.17}{19.26 - 4.9} = \frac{17.8}{14.36}$
   $= 1.239$

(17.8 is the value of the numerator)
(14.36 is the value of the denominator)

Key terms
Brackets
Indices
Operation

Exam tip
Work out the value of the numerator and the denominator first.

Exam-style questions

1 a   Work out $12 - 4 \times 2$ **[1]**
   b   Put brackets where appropriate to make this statement true. $3 + 9 - 5 \times 2 = 11$ **[1]**
2 Here are 5 different symbols $\div$ (+   −   )
   Use each symbol once only to make this statement true.
   7   10   3   2   =   5 **[2]**

3 Use your calculator to work out $\frac{1.5^2 + 3.6}{7.4 - \sqrt{1.44}}$ **[2]**

Exam tip
Check by using the rules of BIDMAS.

CHECKED ANSWERS

Multiplying decimals and negative numbers

Rules

1. To start any calculation, first ignore the decimal points.
2. Carry out the multiplication of whole numbers by your preferred method.
3. Make sure the decimal point in your answer is in the correct place.
4. Negative × negative = positive and positive × negative = negative.

Worked examples

a One metre of wood weighs 3.56 kg.

Work out the weight of 0.6 metres of this wood.

Answer

1. Ignore the decimal points and work out 356×6

2.
$$\begin{array}{r} 356 \\ \times\ \ 6 \\ \hline 2136 \\ {\scriptstyle 33} \end{array}$$

3.
$$3.56 \qquad \times \qquad 0.6 \qquad = \qquad 2.136 \text{ kg}$$

2 decimal places + 1 decimal place = 3 decimal places

b Work out 15.3×1.9

Answer

1. Ignore the decimal points and work out 153×19

2.

	100	50	3	
10	1000	500	30	Adding gives 1530
9	900	450	27	Adding gives 1377 +
				2907

3. $5.3 \times 1.9 = 29.07$ (since the answer is about 30)

c Work out $-2 \times -3 + 4 \times -8$

Answer

4. $-2 \times -3 = (+)6$

4. $4 \times -8 = -32$

so $(+)6 + -32 = -26$

Key terms

Decimal point

Decimal places

Exam tip

Work out an estimate

15.3 is about 15

1.9 is about 2

$15 \times 2 = 30$;

so the answer will be about 30

Exam-style questions

1 Work out 439×1.4 **[3]**

2 Jermaine wants to buy 120 roses as cheaply as possible. Shop A sells 10 roses for £5.36. Shop B sells 6 roses for £3.24.

From which shop should Jermaine buy the 120 roses? **[4]**

3 The electricity readings at the start and the end of a 3-month period were 502.7 kWh and 547.3 kWh. Electricity costs 23.5 p/kWh. Work out total cost of the electricity used in this 3-month period. **[4]**

4 Calculate $(-5 \times 2) + (-3 \times -4)$.

Exam tip

Clearly label your working for shop A or shop B. Write your final answer in a sentence, showing clearly the two total costs **or** the cost of each rose from shop A and shop B.

CHECKED ANSWERS

Dividing decimals and negative numbers

Rules

❶ Rearrange the calculation so that you are dividing by a whole number.
❷ Carry out the division by the whole number by your preferred method.
❸ To get more decimal places in your answer, add zeros to the decimal number you are dividing.
❹ Negative ÷ negative = positive and negative ÷ positive = negative

Worked examples

a The area of a rectangle is 4.252 square metres.

Work out the length of the rectangle if the width is 0.8 metres.

Answer

$4.252 \times 10 = 42.52$

and $0.8 \times 10 = 8$

❶ so $4.252 \div 0.8$ is the same as $42.52 \div 8$

❷
$$\begin{array}{r} 5.3\,1\,5 \\ 8\overline{)42.5^{2}2^{1}4\,0} \end{array}$$

Add a 0 to the remainder of 4 and then divide 40 by 8

so length = 5.315 metres

b Work out $146.4 \div 0.16$

Answer

❶ $146.4 \div 0.16$ is the same as $14\,640 \div 16$

❷
$$\begin{array}{r} 91.5 \\ 16\overline{)14640} \\ -\underline{144} \\ 24 \\ -\underline{16} \\ 80 \\ -\underline{80} \\ 0 \end{array}$$

$1 \times 16 = 16$
$2 \times 16 = 32$
$3 \times 16 = 48$
$4 \times 16 = 64$
$5 \times 16 = 80$
$6 \times 16 = 96$
$7 \times 16 = 112$
$8 \times 16 = 128$
$9 \times 16 = 144$

c Work out $(16 \div -2) + (-15 \div -3)$

Answer

❹ $-8 \div -2 = 4$ and $-6 \div 3 = -2$

so $4 + -2 = 2$

Exam tip

Multiply each number in this case by 10, so that you divide by whole numbers.

Key terms

Decimal point

Decimal places

Exam tip

(to help with long division)

Write down the first 9 multiples of the number you are dividing by (16 in this example).

Exam-style questions

1 Five identical pens cost £39.60

Work out the cost of one pen. **[2]**

2 A rope is 5.32 metres in length.

How many 0.8 metre pieces can be cut from this rope? **[2]**

3 Jane buys a car.
She agrees to pay a deposit of £1500 and 36 equal monthly instalments.
In total she will have to pay £10 629.60

Work out the cost of each monthly instalment. **[4]**

4 Calculate $(16 \div -2) + (-15 \div -3)$

Exam tips

Set out your working clearly.

Check that your answers are realistic.

Using the number system effectively

Rules

In order for the place value of each digit to be correct:
1. When multiplying by 0.1 or 0.01 or 0.001, etc. move the decimal point 1, 2 or 3 places, etc. to the LEFT. This is the same as dividing by 10 or 100 or 1000, etc. The answer will be **smaller** in value.
2. When dividing by 0.1 or 0.01 or 0.001, etc. move the decimal point 1, 2 or 3 places, etc. to the RIGHT. This is the same as multiplying by 10 or 100 or 1000, etc. The answer will be **greater** in value.

Worked examples

a   Work out
   i 34.29×0.1
   ii $34.29 \div 0.1$
   iii 34.29×1000
   iv $34.29 \div 0.0001$

   Answers
   1 i $34.29 \times 0.1 = 3.429$ ← one place to the left
   2 ii $34.29 \div 0.1 = 342.9$ → one place to the right
   2 iii $34.29 \times 1000 = 34290$ ← three places to the right
                                    (one 0 was added to 34.29)
   2 iv $34.29 \div 0.0001 = 342900$ → four places to the right
                                    (two 0s were added to 34.29)

b   Given that $52.4 \times 3.75 = 196.5$, work out
   i 5.24×0.375
   ii $19.65 \div 0.524$

   Answers
   1 $5.24 = 52.4 \div 10$ (or $\times 0.1$)
      $0.375 = 3.75 \div 10$ (or $\times 0.1$)
      So $5.24 \times 0.375 = 196.5 \div 10 \div 10$
                            $= 1.965$
   1 $19.65 = 196.5 \div 10$ (or $\times 0.1$)
      $0.524 = 52.4 \div 100$ (or $\times 0.01$)
      So $19.65 \div 0.524 = \frac{196.5 \div 10}{52.5 \div 100}$
                            $= 3.75 \times 10 = 37.5$

Key term

Place value

Exam tip

Your answer will contain the same digits as the value in the question, e.g. 1965 in **bi** and 375 in **bii**.

Exam-style questions

1 Here is an input/output machine.

   input ⟶ ×0.01 ⟶ ÷10 ⟶ output

   a   Work out the output when the input is 539 **[2]**
   b   Work out the input when the output is 4.58 **[2]**

2 Given that $119 \times 0.35 = 41.65$ work out:
   a 11.9×350 **[1]**
   b $0.4165 \div 0.035$ **[1]**
   c 11.9×0.07 **[2]**

Exam tip

Use inverse operations when given the output.

Look out for

Use the rules above to check that you have not placed decimal points in the wrong place.

CHECKED ANSWERS

Understanding standard form

Rules

To write a number in standard form:
1 Move the decimal point a number of places so that it is immediately after the first non-zero digit and multiply by a suitable power of 10.
2 If the decimal point has been moved to the left, the power of 10 will be plus the number of places moved.
3 If the decimal point has been moved to the right, the power of 10 will be minus the number of places moved.

Key terms

Standard form

Ordinary number

Powers

Worked examples

a Write these numbers in standard form.
  i 347 000
  ii 0.002 18

Answers

i $347\,000 = 347\,000.0 = 3.47 \times 10^?$

5 places to the left

2 means a power of +5

$347\,000 = 3.47 \times 10^5$

ii $0.002\,18 = 2.18 \times 10^?$

3 places to the right

3 means a power of −3

$0.002\,18 = 2.18 \times 10^{-3}$

Exam tips

If the number is large, the power will be positive.

If the number is small, the power will be negative.

b Which number has the greater value, 1.75×10^{-9} or 8.19×10^{-10}?

Answer

$1.75 \times 10^{-9} = 0.000\,000\,001\,75$ (moving the decimal point 9 places to the left (reverse of 3; note 9 zeros))

$8.19 \times 10^{-10} = 0.000\,000\,000\,819$ (moving the decimal point 10 places to the left (reverse of 3; note 10 zeros))

$0.000\,000\,001\,75$ is greater than $0.000\,000\,000\,819$

Exam tip

Change numbers into ordinary numbers by following the rules above in reverse.

Exam-style questions

1 Write these numbers in standard form.
  a 0.072 **[1]**
  b 238.9×10^3 **[1]**

2 Write these numbers as ordinary numbers.
  a 9.14×10^6 **[1]**
  b 5.18×10^{-4} **[1]**

3 Which has the greater value, 167.8×10^{-3} or 17×10^{-2}? **[2]**

Look out for

Numbers that are neither an ordinary number nor a number in standard form, e.g. 238.9×10^3

Exam tip

Make sure that numbers are written in the same format, standard form or as ordinary numbers, before comparing.

CHECKED ANSWERS

Rounding to decimal places and approximating

Rules

❶ To round a number to decimal places, look at the next number after the required number of decimal places; ❶ⓐ if it is 5 or above, increase the previous place number by 1; ❶ⓑ if it is less than 5, do not change the previous place number.

❷ To estimate the approximate answer to a calculation, round each number to **one** significant figure (1 s.f.).

Worked examples

a Write 4.754 correct to 1 decimal place.

Answer
❶ $4.754 = 4.8$

ⓐ the next number after the required number of decimal places

b Write 0.01278 correct to 2 decimal places.

Answer
❶ $0.01278 = 0.01$

ⓑ the next number after the required number of decimal places

c Write down an estimate for the value of
  i   1026
  ii 0.498

Answers
❷ i $1026 \approx 1000$
  ii $0.498 \approx 0.5$

Key terms

Decimal places

Approximation

Estimate

Exam tip

The size of the number does not change.

Look out for

A common mistake is to write 0.498 = 0

Exam-style questions

1 The dimensions of a rectangle are $4.87\,\text{cm} \times 2.35\,\text{cm}$.
Work out the area of this rectangle.
Give your answer correct to 2 decimal places. **[2]**

2 Find an estimate for the value of $\frac{4.83 \times 204}{0.51}$ **[2]**

CHECKED ANSWERS

Multiplying and dividing fractions

Rules

1. When multiplying fractions, multiply the numerators together and multiply the denominators together.
2. When dividing fractions, invert the fraction (turn the fraction upside down) that you are dividing by, then multiply the fractions together using Rule **1**.
3. Always simplify your answer by cancelling.

Worked examples

a Work out

i $\frac{2}{3} \times \frac{5}{7}$

ii $\frac{1}{4} \times \frac{3}{5} \times \frac{2}{9}$

Answers

i **1** $\frac{2}{3} \times \frac{5}{7} = \frac{2 \times 5}{3 \times 7} = \frac{10}{21}$

ii **1** $\frac{1}{4} \times \frac{3}{5} \times \frac{2}{9} = \frac{1 \times 3 \times 2}{4 \times 5 \times 9} = \frac{6}{180}$

3 simplifying $\frac{6}{180}$ gives $\frac{1}{30}$

b Work out

i $\frac{5}{9} \div \frac{1}{3}$ ii $\frac{21}{40} \div \frac{24}{35}$

Give your answers in their simplest form.

Answers

i $\frac{5}{9} \div \frac{1}{3} = \frac{5}{9} \times \frac{3}{1} = \frac{15}{9} = \frac{5}{3}$

　　　　　2　**1**　**3**

ii $\frac{21}{40} \div \frac{35}{24} = \frac{\overset{3}{21}}{\underset{5}{40}} \times \frac{\overset{3}{24}}{\underset{5}{35}}$ **2** $= \frac{3 \times 3}{5 \times 5}$ **1** $= \frac{9}{25}$

Exam tip

If 'giving your answer in its simplest form' is not asked for $\frac{6}{180}$ would get full marks.

Key terms

Numerator
Denominator
Product
Quotient

Exam tip

It is sometimes easier to cancel numbers in the numerator with numbers in the denominator before multiplying out.

Exam-style questions

1 Work out

a $\frac{3}{10} \times \frac{7}{12} \times \frac{5}{42}$ **[2]**

b $\frac{12}{9} \div \frac{18}{30}$ **[2]**

2 Mike, Ali and Emily share some money. Mike has $\frac{2}{3}$ of the money.
Ali has one quarter of the amount that Mike has.
Emily has the rest of the money.
a What fraction of the money does Ali have? **[2]**
Mike's share is divided into 5 equal parts.
b What fraction of the original sum of money are each of these parts? **[2]**

Exam tip

Dividing by 5 is the same as dividing by $\frac{5}{1}$

CHECKED ANSWERS

Adding and subtracting fractions and working with mixed numbers

Rules

1. When adding or subtracting fractions, find equivalent fractions so that all denominators are the same number.
2. Given a mixed number, to convert to a top-heavy or improper fraction, multiply the whole number by the denominator and add the numerator. This gives the new numerator.
3. Given a top-heavy (improper) fraction, to convert to a mixed number, divide the numerator by the denominator to get the whole number. The remainder is then the new numerator of the fraction part of the mixed number.

Worked examples

a Work out

  i $\frac{2}{5}+\frac{1}{6}$

  ii $\frac{7}{8}-\frac{3}{7}$

Answers

i $\frac{2}{5}+\frac{1}{6}=\frac{12}{30}+\frac{5}{30}$ ❶ $=\frac{17}{30}$

ii $\frac{7}{8}-\frac{3}{7}=\frac{49}{56}-\frac{24}{56}$ ❶ $=\frac{25}{56}$

b **i** Work out $\frac{2}{3}+\frac{3}{4}-\frac{1}{5}$

  ii Work out $2\frac{2}{3}\times3\frac{3}{4}$

Give your answers as mixed numbers.

Answers

i $\frac{2}{3}+\frac{3}{4}-\frac{1}{5}=\frac{40}{60}+\frac{45}{60}-\frac{12}{60}$ ❶ $=\frac{73}{60}=1\frac{13}{60}$

❸ $73\div60=1$ remainder 13

ii $2\times3=6$ $6+2=8$

❷ ❷ ❸

$2\frac{2}{3}\times3\frac{3}{4}=\frac{8}{3}\times\frac{15}{4}=\frac{120}{12}=10$

Look out for

Do **not** try to add or subtract fractions if the denominators are different.

Key terms

Numerator

Denominator

Improper fraction

Top-heavy fraction

Mixed number

Exam tip

To find a common denominator, find the LCM (prime factorisation) of all denominators.

Exam-style questions

1 Work out

  a $\frac{5}{8}+\frac{1}{3}$ **[2]**

  b $5\frac{1}{4}-2\frac{5}{12}$ **[2]**

2 A bag contains some counters. $\frac{2}{5}$ of the counters are red; $\frac{3}{8}$ of the counters are blue. The rest of the counters are yellow.

  a What fraction are not blue? **[1]**

  b What fraction are yellow? **[2]**

3 Here is a rectangle.
Work out the area of the rectangle. **[3]**

    $5\frac{1}{3}$ metres

    $2\frac{1}{8}$ metres

Exam tip

It is often easier to convert to improper fractions instead of considering the whole numbers and fractions separately when adding or subtracting mixed numbers.

CHECKED ANSWERS ☐

Converting fractions and decimals to and from percentages

LOW

Rules

❶ To convert a fraction or a decimal to a percentage, multiply the fraction or decimal by 100 and then evaluate.
❷ To convert a percentage to a fraction, write the percentage over 100 and simplify the fraction.
❸ To convert a percentage to a decimal, divide the percentage by 100.

Worked examples

a Convert to %

   i $\frac{3}{8}$

   ii $1\frac{1}{7}$

   iii 0.6

   iv 4.28

   Answers

   i $\frac{3}{8} \times 100$ ❶ $= \frac{300}{8} = 300 \div 8 = 37.5\%$

   ii $1\frac{1}{7} \times 100$ ❶ $= \frac{8}{7} \times 100 = \frac{800}{7} = 800 \div 7 = 114.285...\%$ or $114\frac{2}{7}\%$

   iii 0.6×100 ❶ $= 60\%$

   iv 4.28×100 ❶ $= 428\%$

b Convert
   i   8.5%
   ii 240% to a fraction and a decimal.

   Answers

   i $8.5\% = \frac{8.5}{100} = \frac{17}{200}$ ❷; $8.5\% = 8.5 \div 100 = 0.085$ ❸

   ii $240\% = \frac{240}{100} = \frac{24}{10} = \frac{12}{5} = 2\frac{2}{5}$ ❷; $240\% = 240 \div 100 = 2.40$ ❸

Remember

Percentage means **out of 100**

Key term

Percentage

Look out for

Percentages can also be over 100

Exam tip

Always write down the method; either multiplying by 100 or dividing by 100

Exam-style questions

1 Write the following in order of size. Start with the smallest value. **[2]**

   20% $\frac{2}{9}$ 0.21 $\frac{1}{4}$ 0.202

2 Work out the difference between 85% and $\frac{9}{11}$.

   Give your answer as a decimal. **[2]**

3 There are some pencil crayons in a box. 20% of the crayons are red;

   $\frac{3}{8}$ of the crayons are blue. The rest of the crayons are either

   green or yellow. There are the same number of green crayons as yellow crayons.

   What percentage of yellow crayons are there in the box? **[3]**

Exam tip

It is best to convert all numbers to decimals or percentages; never convert to fractions.

CHECKED ANSWERS

Calculating percentages and applying percentage increases and decreases to amounts

REVISED ☐

MEDIUM

Rules

1a Find the actual increase by finding the percentage of the given amount, then add this value to the original amount.

1b A multiplier can be found by adding the percentage increase to 100 then dividing by 100. The increased amount is then found by applying this multiplier.

2a Find the actual decrease by finding the percentage of the given amount, then subtract this value from the original amount.

2b A multiplier can be found by subtracting the percentage decrease from 100 then dividing by 100. The reduced amount is then found by applying this multiplier.

Key terms

Percentage

Increase

Decrease

Multiplier

Worked examples

a Increase £250 by 12%

Answer

Method 1
1a 12% of 250 = $\frac{12}{100} \times 250 = 30$

250 + 30 = £280

Method 2
1b Multiplier = $\frac{100+12}{100} = \frac{112}{100} = 1.12$

250 × 1.12 = £280

Exam tip

When using multipliers, show how they are found.

b In a sale the cost of a coat is reduced by 35%. Work out the sale price of the coat if it originally cost £79

Answer

Method 1
2a 35% of £78.50 = $\frac{35}{100} \times 79 = 27.65$

79 − 27.65 = £51.35

Method 2
2a Multiplier = $\frac{100-35}{100} = \frac{65}{100} = 0.65$

79 × 0.65 = £51.35

Exam-style questions

1 A ball of string has 8.5 metres in length of string. A piece of string is cut from this ball. The piece of string is 1.5% of the length of string on the ball. Work out the length of this piece of string. Give your answer in centimetres. **[2]**

2 Peter buys a painting for £360. If he sells the painting in an auction he is likely to make a 12.5% profit. Peter sells the painting privately for £400. Could Peter have made a greater profit if he had sold the painting at the auction? **[4]**

3 There are 48 red and 60 white counters in a bag. $33\frac{1}{3}$% of the red counters are removed. The number of white counters is increased by 20%. Are there more or fewer counters now in the bag? **[4]**

Exam tip

Work in consistent units (cm here).

Exam tip

Show clearly each part of your method.

Look out for

A common mistake is to decrease when you should be increasing a value (and vice versa).

CHECKED ANSWERS ☐

Finding the percentage change from one amount to another

Rules

❶ To find one quantity as a percentage of another, write the quantity as a fraction of the other and multiply by 100.

❷ To find percentage change, write the change from the original amount as a fraction and then multiply by 100.

Worked examples

Simplify by cancelling.

a Find 2.4 as a percentage of 15

Answer

$\frac{2.4}{15} \times 100$ ❶ $= \frac{240}{15} = \frac{48}{3} = 16\%$

b The average speed of a train increased from 208 mph to 234 mph. Work out the percentage change.

Answer

$234 - 208 = 26$

% change $= \frac{26}{208} \times 100 =$ ❷ $\frac{2600}{208} = 12.5\%$ increase.

Key term

Percentage

Exam tips

On a calculator paper use your calculator but write down what calculations you are doing.

On a non-calculator paper fractions will usually cancel.

Exam-style questions

1 Work out 65p as a percentage of £26 **[2]**

2 At the start of summer, Sam weighed 80 kg. Over the summer, Sam's weight increased by 2.5%. Sam then went on a diet and has now lost 5 kg in weight.

Work out the percentage change in Sam's weight from the start of summer to now. **[4]**

3 Nazia and Debra are market traders. Nazia paid £428 for some goods and sold them for £492.20. Debra paid £296 for some goods and sold them for £338.92.

Who made the greater percentage profit? **[4]**

Exam tip

Always state if increase or decrease.

Exam tip

You must always explain your answer. An answer of 'Nazia' or 'Debra' gets no credit.

CHECKED ANSWERS

Mixed exam-style questions

1. Paul says $5 \times 2 + 3 - 7$ is equal to 18.

   Lisa says $5 \times 2 + 3 - 7$ is equal to 6.

   By inserting a pair of brackets, show that both Paul and Lisa could be right. **[2]**

2. Naomi is paid £8.45 per hour for a 35-hour week.

   She works 12 hours overtime at a rate of £12.60 per hour.

   Izmail says 'If I work for 45 hours at £9.80 per hour, I will earn more than Naomi.'

   Is Izmail right? **[4]**

3. Sweets are sold in tubes and in boxes.

   A tube contains 40 sweets and costs 48p.

   A box contains 112 sweets and costs £1.40

   a Which offers the better value for money, a tube or a box of sweets? **[3]**

   b Mary needs 180 sweets to decorate a cake.

   What is the most economical way of buying enough sweets? **[2]**

4. Given that $27.3 \times 5.9 = 161.07$, work out

   a 0.0273×59 **[1]**

   b 27.3×5.8 **[2]**

5. The diagram shows a right-angled triangle.

   a Write the dimensions correct to 1 decimal place then work out the area of the triangle. **[3]**

   b If the dimensions were written correct to 2 decimal places, would the area of the triangle be greater than your answer in **a** or less? Explain your answer. Do not work out the actual area to answer this. **[1]**

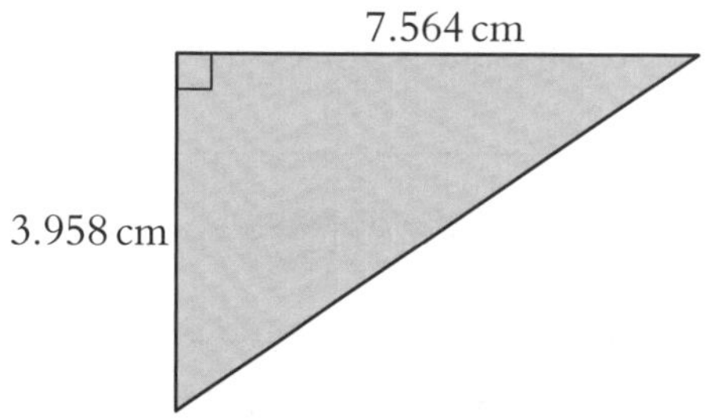

   7.564 cm

   3.958 cm

6. In a cinema there are 29 rows of seats with 41 seats in each row.

   The cost of a ticket for this cinema is £5.95.

   Last night every seat in the cinema was taken.

   Estimate the total cost of the tickets for last night. **[3]**

Sharing in a given ratio

HIGH

Rules

1. To share an amount in the ratio $a : b : c$, find the value of one unit of the amount, divide the amount by the sum of a, b and c.
2. Then multiply this answer by each of a, b and c.
3. To find the fraction of each part of the ratio $a : b : c$, write each part as a fraction out of the sum of a, b and c.

Worked examples

a Tom, Mary and Sally share £72 in the ratio $4 : 3 : 2$.
Work out how much each person gets.

Answer

1 $72 \div (4 + 3 + 2) = 72 \div 9 = £8$ per share

2 Tom gets £8 × 4 = £32,
Mary gets £8 × 3 = £24,
Sally gets £8 × 2 = £16

b The ratio of a mixture of cement and sand is $1 : 4$.
What fraction of the mixture is
  i cement
  ii sand?

Answers

3 i cement $= \frac{1}{1+4} = \frac{1}{5}$

  ii sand $= \frac{4}{1+4} = \frac{4}{5}$

Key terms

Ratio

Proportion

Look out for

A common mistake is to write $\frac{1}{4}$ for the fraction of cement.

Exam-style questions

1 Divide £132 in the ratio $5 : 4 : 2$ **[2]**

2 In an election, the ratio of Conservative voters to Labour voters to other voters is $9 : 5 : 3$. 27 132 people voted in this election.

How many more people voted Conservative than Labour? **[3]**

3 A shop sells coffee in three different sizes of packet. The volume of coffee in each packet is in the ratio $3 : 4 : 5$. The cost of each size of packet is £4.30, £6.30 and £7.50.

Which size of packet gives the best value for money? **[4]**

Exam tips

Always work out the value of **one** share.

Always check that your answers add up to the amount given.

Exam tip

Make sure your choice is supported by clear working.

CHECKED ANSWERS

Working with proportional quantities

HIGH

Rules

❶ To use the unitary method, find out what proportion is just **one** part of the whole amount,

❷ Then multiples of that can be found.

Key terms
Ratio
Proportion
Multiples

Worked examples

a 12 identical books cost £23.88.
Work out the cost of 5 of these books.

Answer
❶ 23.88 ÷ 12 = £1.99 per book ⟵———————— The value of **one** unit.
❷ 5 books cost £1.99 × 5 = £9.95

b Work out which is the better value for these bags of potatoes;
6 kg for £8.16 or 11 kg for £15.18.

Answer
❶ £8.16 ÷ 6 = £1.36 per kg

❶ £15.18 ÷ 11 = £1.38 per kg

So 6 kg for £8.16 is the better value.

Exam tip
Always give answers in a sentence supported by working.

Exam-style questions

1 8 pens cost £5.20.

Work out the cost of 15 of these pens. **[2]**

2 Jay buys three portions of chips and two pies for £6.45. Mandy buys five pies for £6.

How much does one portion of chips cost? **[3]**

3 Here are the ingredients to make 40 biscuits.
600 g of butter, 300 g of sugar and 900 g of flour.
Mrs Bee has the following ingredients in her cupboard.
1.5 kg of butter, 1 kg of sugar and 2 kg of flour.

Work out the greatest number of these biscuits that Mrs Bee can make. **[4]**

Exam tip
Always work out the value of **one** part.

Exam tip
Explain why this is the greatest number.

CHECKED ANSWERS

Index notation

HIGH

Rules

❶ $a \times a \times a \times ... \times a$ (m times) is written a^m.

❷ To multiply numbers written in index form, add the powers together. $a^m \times a^n = a^{m+n}$

❸ To divide numbers written in index form, subtract the powers. $a^m \div a^n = a^{m-n}$

❹ To raise a number written in index form to a given power, multiply the powers together. $(a^m)^n = a^{mn}$

Key terms

Index

Indices

Powers

Worked examples

a Write $7 \times 7 \times 7 \times 7 \times 7$ in index form.

Answer

❶ $7 \times 7 \times 7 \times 7 \times 7 = 7^5$

b Write $\left(\frac{2^3 \times 2^4}{2^5}\right)^3$ as a power of 2.

Answer

$\left(\frac{2^3 \times 2^4}{2^5}\right)^3 = \left(\frac{2^7}{2^5}\right)^3$ since $2^3 \times 2^4 = 2^{3+4} = 2^7$ ❷

$= (2^2)^3$ since $2^7 \div 2^5 = 2^{7-5} = 2^2$ ❸

$= 2^6$ since $(2^2)^3 = 2^{2 \times 3} = 2^6$ ❹

Exam tip

A common mistake is to multiply powers instead of adding in a product.

Exam tip

A common mistake is to divide powers instead of subtracting in a quotient.

Look out for

Follow the rules of BIDMAS and work out the calculation inside the brackets first.

Exam-style questions

1 a Write $10 \times 10 \times 10 \times 10$ in index notation. **[1]**
  b Use your calculator to work out the value of 8^5 **[1]**

2 $x = 8 \times 2^4$ $y = 4^2 \times 16$
  Work out the value of xy. Give your answer as a power of 2 **[3]**

3 Tom is trying to work out the value of $\frac{10^4 \times 10^5}{10 \times 10^2}$

  Tom writes $\frac{10^4 \times 10^5}{10 \times 10^2} = \frac{10^{20}}{10^2} = 10^{10} = 100$.

  Write down each of the mistakes that Tom has made. **[4]**

CHECKED ANSWERS

Prime factorisation

Rules

To write a number as a product of its prime factors, use either the factor-tree method or the method of repeated division.
- **①** Factor tree method: continue to write each number as a product of two factors until all of the factors are prime numbers; then write these as a product.
- **②** Repeated division method: continue to divide by a prime number until the final answer is 1; then write as a product all of the prime numbers that have been used.
- **③** To find the HCF of numbers written as products of their prime factors, choose all common factors and multiply together.
- **④** To find the LCM of numbers written as products of their prime factors, choose all factors (common factors just once) and multiply together.

Key term

Prime factors

Worked examples

a Write 108 as a product of its prime factors.

Answer

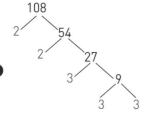

2	108
2	54
3	27
3	9
3	3
	1

$108 = 2 \times 2 \times 3 \times 3 \times 3$

① ...

② $108 = 2 \times 2 \times 3 \times 3 \times 3$

Exam tip

This answer can also be written in index notation as $2^2 \times 3^3$

Look out for

A common mistake is to list the prime factors without writing as a product: 2, 2, 3, 3, 3 would lose marks.

b Find   **i** the HCF of 108 and 80   **ii** the LCM of 108 and 80

Answers

108 written as a product of its prime factors is $2 \times 2 \times 3 \times 3 \times 3$

80 written as a product of its prime factors is $2 \times 2 \times 2 \times 2 \times 5$ using **①** or **②**.

i The HCF of 108 and 80 = $2 \times 2 = 4$ **③** choosing common prime factors.

ii The LCM of 108 and 80 = $2 \times 2 \times 3 \times 3 \times 3 \times 2 \times 2 \times 5 = 2160$ **④** choosing all prime factors but writing common factors once only.

Exam tip

The LCM is simply the HCF multiplied by all other factors.

Exam-style questions

1 a Write 96 as a product of its prime factors. **[2]**

  b Find   **i** the HCF of 96 and 120   **ii** the LCM of 96 and 120 **[3]**

2 Nadir is in hospital. She has an injection every 6 hours. She has tablets every 8 hours. She has her bandages changed every 10 hours.
On Monday at 8.00 am Nadir is given an injection, she takes her tablets and has her bandages changed.

When will all three treatments be given to her next at the same time? **[3]**

CHECKED ANSWERS ☐

Mixed exam-style questions

1. $\frac{5}{6}$ of the members of a sports club are male. $\frac{1}{4}$ of the male members are under 18 years of age. $\frac{1}{3}$ of the female members attend keep fit classes.

    a What fraction of the members of the sports club are male and over 18? [2]

    b What fraction of the members of the sports club are female and do not attend keep fit classes? [2]

2. The table shows the time spent by Helen doing homework last week.

Day	Mon	Tues	Wed	Thurs	Fri
Time in hours	$1\frac{3}{4}$	$2\frac{1}{2}$	$1\frac{1}{3}$	$3\frac{2}{7}$	$\frac{3}{4}$

   $\frac{1}{3}$ of her time spent doing homework was in either Maths or English.

   Work out the number of hours spent doing either Maths or English. Give your answer as a mixed number. [3]

3. Rachel's annual salary is £34 500. Peter's annual salary is £32 900.
   Rachel gets a 4% increase in her salary.
   Peter gets a 6% increase in his salary.
   Whose annual salary is now the greater? [3]

4. In 2015, Jason had 5800 stamps in his collection.
   In 2016, Jason sold some stamps and reduced his collection by 15%.
   In 2017, Jason sold some more stamps reducing his collection by a further 10%.

    a In 2016 and 2017, how many stamps did Jason sell altogether? [3]

    b What was the overall percentage reduction in Jason's collection of stamps? [2]

5. Brian spends $\frac{2}{5}$ of his monthly earnings on clothes and entertainment.
   The ratio of money spent on clothes to money spent on entertainment is $4:3$.
   What fraction of his monthly earnings does Brian spend on entertainment? [3]

6. Concrete is made from sand, stone and cement.
   Bill makes 10 cubic metres of concrete with the ratio of sand, stone and cement equal to $4:3:1$.
   Sandra makes 8 cubic metres of concrete with the ratio of sand, stone and cement equal to $6:5:2$.

    a Who uses the most cement, Bill or Sandra? [3]

    b If Bill mixed his concrete with sand, stone and cement in the ratio $10:7:2$, how would this affect the amount of cement that he would need? [1]

7. Milk is sold in three sizes of bottle, small, medium and large.
   A small bottle holds 1 pint of milk and costs £1.24.
   A medium bottle holds 1 litre of milk and costs £2.15.
   A large bottle holds 1 gallon of milk and costs £9.80.
   Which size bottle of milk is the better buy? (1 pint = 0.568 litres) [4]

Algebra: pre-revision check

Check how well you know each topic by answering these questions. If you get a question wrong, go to the page number in brackets to revise that topic.

1 This formula gives the value of p in terms of q and r.

$p = 2q - 3r$.

Find the value of p when $q = 10$ and $r = 4$.   (page 20)

2 Solve these equations.
  a $a + 4 = 6$
  b $\frac{b}{3} = 5$
  c $5c + 4 = 6$
  d $15 - \frac{3e}{2} = 24$ (page 21)

3 a Expand these brackets.
  i $5(2a + 3)$
  ii $h(3h - 6)$
  iii $3x(4x - 2y)$
  b Factorise fully
  i $6y + 12$
  ii $6p^2 - 9p$
  iii $5e^2 + 10ef$
  iv $8x^2y - 12xy^2$ (page 22)

4 Solve these equations.
  a $5x - 6 = 2x + 3$
  b $7 - 2p = 6p + 13$
  c $2 - \frac{3y}{2} = 5 + \frac{5y}{4}$ (page 23)

5 Solve
  a $5(3g - 2) = 35$
  b $4(5h + 7) = 3(2h + 8)$
  c $2(5k + 8) - 6 = 4(2k - 1)$ (page 24)

6 Here are the first 5 terms of a linear sequence.

    4   10   16   22   28

  a Find the nth term of the sequence.
  b Work out the 50th term in the sequence.
  c Explain if 900 is a term in the sequence.
  (page 25)

7 a On a coordinate grid drawn with values of x from -3 to $+3$ and values of y from -6 to $+8$, draw the graph of $y = 2x + 1$.
  b Find the value of x when $y = 6$ (page 27)

8 Here is the graph that shows the depth of water in a harbour. A ship needs to enter the harbour between 08:00 and 20:00. It needs a 4 metre depth of water in the harbour. Between what times can the ship enter the harbour?   (page 29)

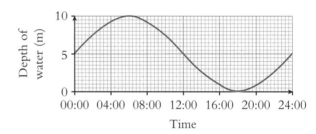

Working with formulae

Rules

❶ You can replace words or letters in a formula with numbers.
❷ Use BIDMAS to find the value of the missing word or letter.
❸ Use inverses to write the formula or equation so that the missing letter is on its own on one side of the formula or equation.

Worked examples

a Here is a formula to find the perimeter of a rectangle:
$P = 2l + 2w$.
Find the value of P when $l = 6$ and $w = 4$.

Answer
$P = 2l + 2w$

❶ $P = 2 \times 6 + 2 \times 4$

❷ $P = 12 + 8 = 20$

b Tom hires a car from Cars 2U.

Cars 2U

i How much does it cost to hire a car for 7 days?

ii Ben has £100.
For how many days can he hire a car?
You must explain your answer.

£20 plus £30 a day

Answers

i Cost = $20 + 7 \times 30$ ❶
Cost = $20 + 210$ ❷
Cost = £230

ii $100 = 20 + N \times 30$ ❶
$100 - 20 = N \times 30$ ❸
$80 = N \times 30$ so $N = 80 \div 30 = 2.666...$ ❷
Ben can hire the car for 2 days.
This costs £80.
3 days cost £110 which is too much.

Look out for

$2w$ means $2 \times w$

So if $w = 5$ then $2w$ is $2 \times 5 = 10$ and not 25

Key terms

Formula

Substitute

Variable

Equation

Exam tips

Always show your working when answering algebra questions.

Do not use trial and error methods as you may well lose marks.

Exam-style questions

1 Bobbie uses this number machine to work out the number of cartons of orange juice she needs for a party.

Number of people → [÷5] → [+2] → Number of cartons

a How many cartons will Bobbie need for 40 people? **[2]**
b How many people are in a party that uses 20 cartons? **[2]**

2 This formula gives the time taken, T minutes, to cook a chicken of weight w kg.
$T = 40w + 20$
a How long does it take to cook a chicken of weight 2.5 kg? **[2]**

It takes 3 hours 20 minutes to cook a different chicken.
b How heavy was the chicken? **[2]**

CHECKED ANSWERS

Setting up and solving simple equations

Rules

1. Always use the inverse operations to solve an equation.
2. + and − are the inverse of one another.
3. × and ÷ are the inverse of one another.
4. To set up an equation a variable must be defined.

Worked examples

a **2** $2p + 5 = 17$ (−5 is the inverse operation of + 5)

   $2p + 5 - 5 = 17 - 5$ (subtract 5 from each side
                                    of the equation)

   3 $2p = 12$ (÷ 2 is the inverse of × 2)

   $p = 6$ (divide each side of the equation by 2)

b **2** Ann is two years younger than Ben.
      Clara is twice as old as Ben.
      The total of their ages is 58.
      Work out their ages.

   Answer
   Let Ben's age be x

   Ann's age will be $x - 2$

   Clara's age will be $2x$

   Firstly, set up the equation:

   $x + (x - 2) + 2x = 58$

   Now collect like terms:

   $4x - 2 = 62$

   $4x = 60$ (Add 2 to each side)

   $x = 15$ (Divide each side by 4)

   Ann is 13, Ben 15 and Clara 30.

Key terms

Equation

Inverse operation

Solve

Variable

Exam tips

Always use algebraic methods and show your working to gain full marks.

Always check your answer to make sure it is correct.

Exam-style questions

1 Solve these equations.
   a $a - 3 = 7$ **[1]**   b $\frac{b}{5} = 3$ **[1]**   c $3c + 9 = 7$ **[2]**
   d $\frac{5d}{2} + 4 = 29$ **[2]**   e $6 - 2e = 3$ **[2]**

2 Here is a rectangle.
   The length is $2x + 5$.
   The width is $x - 3$.
   The perimeter is 46 cm.

   $2x + 5$

   $x - 3$

   Work out the area of the rectangle in cm². **[4]**

Using brackets

Rules

1. When you expand a bracket you multiply what is inside the bracket by the number or variable outside the bracket.
2. When you factorise an algebraic expression you take out the common factor from each term of the expression and put it outside the bracket.

Worked examples

a Expand

   i $4(3x + 5)$ **ii** $t(3t - 4)$

Answers

$4 \times (3x + 5)$

$4 \times 3x + 4 \times 5$

$12x + 20$

$t \times (3t - 4)$

$t \times 3t - t \times 4$

$3t^2 - 4t$

b Factorise

   i $4p + 6$ **ii** $6p^2q - 9pq^2$

Answers

$2 \times 2 \times p + 2 \times 3$
2 is in 4 and in 6
2. $2(2p + 3)$

$3 \times p \times q \times 2 \times p - 3 \times p \times q \times 3 \times q$
3 and p and q are in both terms
2. $3pq(2p - 3q)$

Look out for

$x \times x = x^2$ using index laws.

Key terms

Brackets

Variable

Expression

Expand

Factorise

Exam-style questions

1 Beth is 3 years older than Amy.
Cath is twice as old as Beth.
The total of their ages is 41.
How old are the three girls? **[3]**

2 PQRS is a rectangle.
The length of the rectangle is
$(2x - 5)$ cm.
The width of the rectangle is 6 cm.
The area of the rectangle is 72 cm².
Find the value of x and the perimeter
of the rectangle. **[4]**

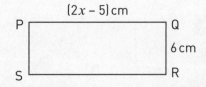

CHECKED ANSWERS

More complex equations and solving equations with the unknown on both sides

REVISED ☐

HIGH

Rules

1. Keep the variables on the side of the equation that has the highest number of that variable and move the numbers to the other side of the equation.
2. Then solve the equation.

Worked examples

a Solve $5x + 4 = 2x - 8$

Answer

$5x - 2x + 4 = -8$ ❶

$3x = -8 - 4$ ❷

$3x = -12$

$x = -4$

b Solve $5.5 - 2.4y = 1.6y - 2.5$

Answer

$5.5 = 1.6y + 2.4y - 2.5$ ❶

$5.5 + 2.5 = 4y$ ❷

$8 = 4y$

$y = 2$

Look out for

The variable on both sides of the equation.

($5x$ is bigger than $2x$ so keep the xs on the left-hand side of the equation)

($1.6y$ is bigger than $-2.4y$ so keep the ys on the right-hand side of the equation)

Key terms

Variable

Solve

Exam-style questions

1 Here is a rectangle.
All measurements are in cm.

Find the area of the rectangle in cm². **[5]**

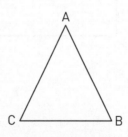

$3r + 4$

$2p + 3$ $4p - 2$

$7r - 8$

2 Here is a triangle.
Angle A = $2x + 30$
Angle B = $5x - 15$
Angle C = $3x + 15$

Prove that triangle ABC is an equilateral triangle.
What assumptions have you made in your proof? **[5]**

A

C   B

Exam tip

Always set up your equations first.

CHECKED ANSWERS ☐

Algebra

Solving equations with brackets

Rules

❶ Multiply out (expand) the brackets.
❷ Then solve the equation.

Worked examples

a Solve $3(2x + 5) = 27$

Answer

$3 \times 2x + 3 \times 5 = 27$ \qquad (Expand the bracket) ❶

$\quad 6x + 15 = 27$ ❷

$\qquad 6x = 12$

$\qquad x = 2$

b Solve $5(3y - 2) = 3(6y + 5) - 13$

Answer

$5 \times 3y - 5 \times 2 = 3 \times 6y + 3 \times 5 - 13$ \quad (Expand the brackets) ❶

$\quad 15y - 10 = 18y + 15 - 13$ ❷

$\quad 15y - 10 = 18y + 2$

$\qquad -12 = 3y$ ←

$\qquad y = -4$

Look out for

Brackets that will need to be multiplied out.

Key terms

Variable
Expand
Bracket
Solve

(Move the variables to the right-hand side as it is the one with the highest number of the variable)

Exam-style questions

1 Abbi thinks of a number n.
She doubles the number and adds 5.
Abbi then multiplies her answer by 5 and gets 85.
Find the number Abbi first thought of. **[3]**

2 Here is a rectangle.
The perimeter of the rectangle
is 210 cm.
Work out the area of the
rectangle. **[3]**

$3(2x + 5)$ cm

$2(2x - 5)$ cm

Exam tip

Always set up your equations first.

CHECKED ANSWERS

Linear sequences

Rules

1. The difference is found by subtracting consecutive terms.
2. If the difference between each term is always the same (common difference) then the sequence can always be written as $an + b$, this is the nth term.
3. The value of a is always the common difference.
4. Use the first term to find the value of b.
5. To check if a number is in a sequence then make an equation with the nth term.
6. By putting whole numbers (1, 2, 3...) into the nth term you can build up the sequence.

Worked examples

a Here is a number pattern: 4   10   16   22 ...
  i   Find the next term in the pattern.
  ii Find the nth term in the pattern.
  iii Explain why 102 is not a member of the pattern.

  Answers
  i   Common difference is 6 so next term is 22 + 6 = 28 ❶
  ii nth term will be $6n + b$ ❷ ❸
      First term is 4 so $(6 \times 1) + b = 4$ so $b = 4 - 6 = -2$ ❹
      nth term is therefore $6n - 2$
  iii If 102 is in the sequence then $6n - 2 = 102$ ❺
      Add 2 to each side gives $6n = 104$
      Divide each side by 6 gives $n = 17\frac{1}{3}$
      For 102 to be in the series then the value of n must be a whole number because n is the term number.
      Therefore, 102 is not in the sequence.

b The nth terms of two linear sequences are: $5n + 2$ and $30 - 6n$
  Explain if these two sequences have any terms that are common.

  Answer
  List the terms in the sequences ❻:

$5n + 2$ gives	7	12	17	22	27	32
$30 - 6n$ gives	24	18	12	6	0	-6

  12 is in both sequences.

Look out for

Always check the difference between consecutive terms.

Key terms

Term

Number sequence

Number pattern

Number series

Difference

Common difference

nth term

Exam-style questions

1 Here are some terms in a linear sequence.   7, ..., 15, ..., ..., t

  a Find the value of term t. **[1]**
  b Find the nth term in the sequence **[2]**
  c Explain if 163 is in the sequence. **[2]**

2 For what value of n does the nth term of this linear sequence first become negative?

  55   51   47   43 ...

  You must show all your working. **[3]**

Exam tip

Always check your answer by listing the terms of the sequence.

CHECKED ANSWERS

Mixed exam-style questions

1   Nigel uses this formula to change between degrees Fahrenheit and degrees Celsius.

   $$C = \frac{5F - 160}{9}$$

   The average normal temperature of a human body is 98.6 °F.
   a   What is 98.6 °F in degrees Celsius?   [2]
   b   What temperature is the same in degrees Celsius as it is in degrees Fahrenheit?   [3]

2   Bobbi uses this formula to work out the time, t minutes, it takes to cook a chicken of weight w kg.
      $t = 40w + 20$
   Bobbi wants a chicken weighing 2 kg to be cooked at 12 noon. At what time should she put
   the chicken into the oven?   [3]

3   A square has a perimeter of $(40x + 60)$ cm. A regular pentagon has the same perimeter as the square.
   Show that the difference between the length of the sides of the two shapes is $(2x + 3)$ cm.   [3]

4   Mrs Jones organises a school trip to the theatre for 42 people.
   She takes children and adults on the trip.
   Each adult paid £40 for their ticket.
   Each child paid £16 for their ticket.
   The total cost of the tickets was £1080.
   How many adults went on the trip?   [4]

5   Here is a T shape drawn on part of a 10 by 10 grid. The shaded
   T is called T_2 because 2 is the smallest number in the T.
   T_2 is the sum of all the numbers in the T shape; so $T_2 = 45$.
   a   Find an expression, in terms of n, for T_n [3]
   b   Explain why T_n cannot equal 130.   [2]

1	2	3	4	5	6
11	12	13	14	15	16
21	22	23	24	25	26
31	32	33	34	35	36
41	42	43	44	45	46

Plotting graphs of linear functions

Rules

1. Always draw up a table of values to help plot the points on the grid.
2. Start with the value 0 and put in positive values first.
3. Plot the points and join with a straight line.
4. You can use the graph to read off values from one axis to the other.

Worked examples

a The time it takes to cook a chicken is given by the formula $T = 20w + 20$
  i Draw a table of values for $T = 20w + 20$
  ii Draw the graph of T for values of w from 0 to 5 pounds in weight (w).
  iii Use your graph to find the time (T) it takes to cook a $3\frac{1}{2}$ pound chicken.

Answers

i ① ②

w	0	1	2	3	4	5
20w	0	20	40	60	80	100
20	20	20	20	20	20	20
$T = 20w + 20$	20	40	60	80	100	120

ii Draw graph. ③

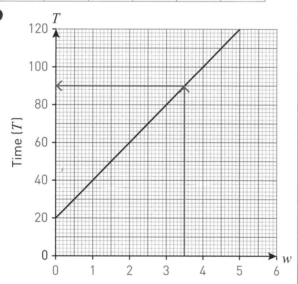

iii Draw a vertical line up from $3\frac{1}{2}$ on the w axis to the graph and then from the graph draw a horizontal line to the T axis. ④
    This gives an answer of 90 mins.

b Here is a table of values for $y = 2x + 1$.

x	−3	−2	−1	0	1	2	3
y							

  i Copy and complete the table of values.
  ii Draw the graph of $y = 2x + 1$.
  iii Find the value of x when $y = 4$.

Look out for

Always work out the positive values first in a table of values.

Make sure the line is straight when you draw it.

Key terms

Variable

Table of values

Axis

Exam tip

Always show how you read off from the graph by drawing in the straight lines.

→

Answers

i   First work out the positive values of *y*. Then work out the negative values. (They should follow the same pattern.) ❶ ❷

x	–3	–2	–1	0	1	2	3
y	–5	–3	–1	1	3	5	7

ii Graph drawn. ❸

iii *x* = 1.5 when *y* = 4 ❹

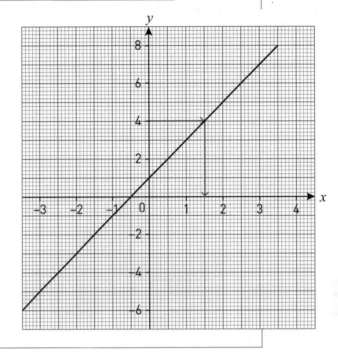

1 Liam hires a car. He pays £40 for the first day then £20 for each extra day.
   a   Draw the graph of the cost of hiring the car. **[2]**
   b   Liam has £150 to spend on car hire. For how many days can Liam hire the car? **[2]**

2 Here is a table of values for *y* = 2*x* – 1

x	–3	–2	–1	0	1	2	3
y							

   a   Copy and complete the table of values. **[2]**
   b   Draw the graph of *y* = 2*x* – 1. **[3]**
   c   Find the value of *x* when *y* = 4. **[2]**

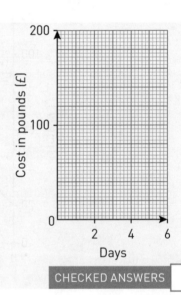

CHECKED ANSWERS ☐

Real-life graphs

Rules

1. The maximum value is the highest point on the graph.
2. The minimum value is the lowest point on the graph.
3. The steeper the line on a graph means the greater the rate of change.
4. The less steep the line on a graph means the lower the rate of change.
5. If the line is horizontal the rate of change is zero.

Worked example

Here are four graphs.

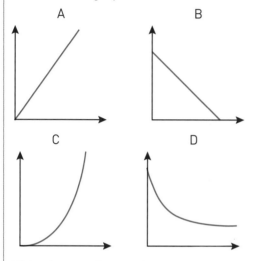

Write down a situation that each of the graphs could describe.

Answer

A As one variable goes up so does the other. These graphs can be used for conversion graphs and where the more you buy the more you pay.

B As one variable goes up the other variable goes down. It can show the fuel left in a car's fuel tank as it goes on a journey.

C As one variable goes up the other variable goes up faster. It can show an increase in population e.g. rabbits. 3

D As one variable goes up the other variable goes down but less slowly as it goes down. It can show the temperature of a cup of tea when left to cool. 4

Look out for

Distance-time graphs and conversion graphs.

Key terms

Variable

Axes

Relationship

Maximum

Minimum

Rate of change

Gradient

Exam tip

Use sentences in worded answers.

Exam-style question

Here is a travel graph of Jo's trip to the shops and back. She had to stop at some roadworks on her way to the shops.

a What time did Jo leave home? **[1]**
b How much time did Jo have to spend at the roadworks? **[1]**
c How much time did she spend at the shops? **[1]**
d At what time did Jo get home again? **[1]**
e How far is Jo's home from the shops? **[1]**
f How much time did it take Jo to get to the shops? **[1]**
g How much time did it take Jo to get home from the shops? **[1]**
h Work out Jo's average speed from the shops to home. **[2]**

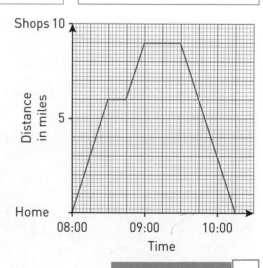

CHECKED ANSWERS

Mixed exam-style questions

1 Sid hires a car from **Cars 4 U**.
   a Find the cost of hiring a car for 4 days
      from **Cars 4 U**.
   b The cost of hiring a car from **Cars 4 U** is £20
      plus a daily rate. Work out the daily rate.
   Sid wants to compare the cost of hiring a car from
   Cars 4 U and from **Car Co** who charge £25 for
   each day of hire. Sid hires cars for different periods
   of time. He wants to use the cheaper company.
   c Which of these two companies is the cheaper to
      hire the car from? You must show your working
      and explain your answer.

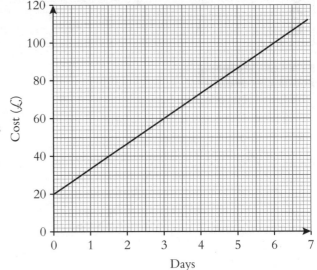

2 On a coordinate grid with values of x from -3 to $+3$
   and values of y from -6 to $+8$, draw the graph of
   $y = 2x + 3$

Geometry and Measures: pre-revision check

Check how well you know each topic by answering these questions. If you get a question wrong, go to the page number in brackets to revise that topic.

1 Approximately how many miles is equivalent to 160 km? (page 33)

2 The bearing of A from B is 235°.
  a Draw the bearing of A from B.
  b Use your diagram to work out the bearing of B from A. (page 34)

3 A plan of a garden design is drawn to a scale of 1 : 25. The total length of fencing shown on the plan is 68 cm.
  a How many metres of fencing will be needed for the actual garden?
The actual lawn will be 3.5 m wide.
  b How wide will the lawn be on the diagram? (page 34)

4 A slug moves at a speed of 1.2 cm/sec. How far will the slug have moved in 2 hours? Give your answer in metres. (page 35)

5 Name the quadrilaterals that have
  a one pair of opposite sides parallel
  b four equal sides
  c rotational symmetry order 2. (page 36)

6 Write down the size of angles a and b. (page 38)

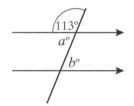

7 Work out the size of the exterior and interior angles of a regular 12-sided polygon. (page 39)

8 Calculate the area and the perimeter of the parallelogram shown below. Give the units of your answers. (page 40)

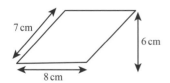

9 Find the area and circumference of a circle with a radius of 5 cm. (page 41)

10 Using a ruler and pair of compasses only, construct the triangle XYZ such that XY = 8 cm, YZ = 6.5 cm and angle XYZ = 90°. (page 43)

11 Triangle A is drawn on the square grid below. (page 45)

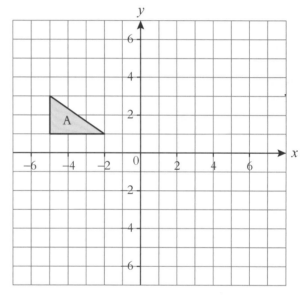

  a Rotate triangle A through 180° about the point with coordinates (0, 2). Label your answer B.
  b Reflect triangle A in the line $y = -1.5$. Label the image C.

12 Describe fully the transformation that maps triangle T to triangle R in the diagram below.   (page 49)

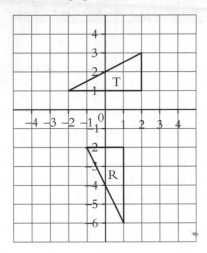

13 Sketch the nets, plan view and side elevation of
 a   a cuboid
 b   a right-angled triangular prism.   (page 51)

14 Calculate the volume and surface area of this cuboid.   (page 52)

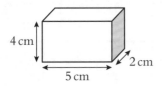

4 cm
2 cm
5 cm

15 Calculate the volume and surface area of this cylinder.   (page 52)

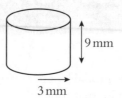

9 mm
3 mm

16 Draw the plan, front and side elevations of the object shown below.   (page 51)

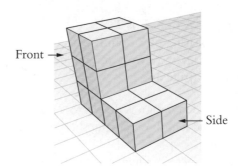

Front
Side

Converting approximately between metric and imperial units

REVISED

LOW

Key terms

Approximate

Estimate

Proportion

Unit

Worked examples

a Trefor travels at a constant rate.
He travels 120 km in 1 hour.
How far will he travel in $\frac{1}{4}$ hour?
Give your answer in **miles**.

Answer
In $\frac{1}{4}$ hour Trefor will travel $120 \div 4 = 30$ km.
8 km ≈ 5 miles
1 km $\approx \frac{5}{8}$ miles (dividing both sides by 8, to find 1 km)
We need to find 30 km in miles.
As 1 km $\approx \frac{5}{8}$ miles,
then 30 km $\approx 30 \times \frac{5}{8}$ (multiplying both sides by 30).
30 km ≈ 18.75 miles
Trefor travels 18.75 miles. (As a check, 1 kilometre is shorter than 1 mile, so we did expect the answer to be a smaller value than 30.)

b There are 12 inches in 1 foot.
How many metres are there in 8 feet?

Answer
Use a known fact: 1 inch ≈ 2.5 cm.
So 12 inches $\approx 12 \times 2.5$ cm. (multiplying by 12)
12 inches ≈ 30 cm
So 1 foot, which is 12 inches, is approximately 30 cm.
8 feet will be $8 \times 30 = 240$ cm.
As there are 100 cm in 1 metre, so 240 cm $\div 100 = 2.4$ metres.
8 feet is approximately 2.4 metres.

Exam-style questions

1 Given that 1 gallon is 8 pints, use the fact that 1 gallon is approximately 5 litres to estimate the number of pints in 300 litres.

2 Approximately how many kilometres are there in 345 miles?

CHECKED ANSWERS

Exam tips

Use proportion, but think if each stage of your working is reasonable.

Always give units to the stages of your work.

Bearings and scale drawings

Rules

1. Bearings are always measured from North in a clockwise direction.
2. Bearings are always given using three figures.
3. A scale drawing is the same shape as the original and all its lengths are in the same ratio.
4. A scale factor is the ratio of the lengths of the original to those in the scale drawing.

Worked examples

a B is 4 cm from A on a bearing of 100°. Draw the bearing of B from A.

Answer

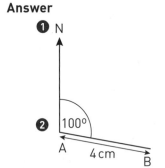

b Daniel has a model of a house on a scale of $\frac{1}{40}$ th. The front of the model house is 0.5 m wide. The model house is 40 cm high.
  i   How wide is the real house?
  ii How high is the real house?

Answer
The real house is 40 times bigger than the model.

  i   Width of real house = 40 × 0.5 m = 20 m ❹
  ii Height of real house = 40 × 40 cm = 1600 cm or 16 m

Key terms

Bearing

Scale drawing

Scale factor

Ratio

Look out for

Make sure you give units with your answer.

To change cm to metres divide by 100.

Exam-style questions

1 Greenfield is a village 6 miles due East of Blackford. The village of Redham is 4 miles from Blackford and lies on a bearing of 135°.
   a   Draw a scale diagram showing the positions of the three villages. **[4]**
   b   Use your diagram to work out the bearing of Redham from Greenfield. **[2]**

2 A map has a scale of 1 : 25 000. On the map the distance between two bridges across a river is 8 cm.
   a   What is the actual distance between the two bridges? **[1]**
   b   The actual length of the river is 6.7 km. How long is the river on the map? **[1]**

3 The bearing of B from A is 080°. Work out the bearing of A from B. **[1]**

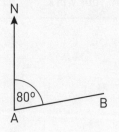

Exam tip

Draw a North line before you start.

Exam tips

Use Rule ❸.

Check your answer is realistic.

CHECKED ANSWERS

Compound units

Rules

❶ A compound measure involves two quantities, for example speed is measured in distance and time.

❷ In a compound unit 'per' means 'for each' or 'for every'.

❸ If you need to change the units of a compound unit change one quantity at a time.

Key terms

Speed

Per

Rate

Worked examples

a Galston runs 300 m in 30 s. Work out his average speed in

  i metres per second

  ii km per min.

Answer

$$\text{average speed} = \frac{\text{distance}}{\text{time}} \quad ❶$$

  i Galston's average speed $= \frac{300\,\text{m}}{30\,\text{s}}$

$$= 10 \text{ metres per second}$$

  ii $10 \times 60 = 600$ metres per minute **❸**

$$= 600 \div 1000 \text{ km per min} = 0.6 \text{ km per min}$$

b A car uses petrol at the rate of 8 km per litre.
How much petrol would the car use for a journey of 300 km?

Answer

Amount of petrol used $= \frac{300}{8} = 37.5$ litres **❷**

Look out for

To change m to km divide by 1000.

To change seconds to minutes divide by 60.

You can use '/' instead of per.

Remember

Don't forget to include units in your answer.

Exam-style questions

1 The distance from London to Larnaca in Cyprus is 3212 km. A plane takes 3 hours and 30 mins to fly from London to Larnaca.

  Work out the average speed of the plane. **[2]**

2 There are 50 litres of water in a barrel. The water flows out of the barrel at a rate of 125 millilitres per second (1 litre = 1000 millilitres).

  Work out the time taken for the barrel to empty completely. **[3]**

3 Sarab and Julian both drive their own cars from London to Leeds. Sarab's car averages 10 km per litre of petrol, and uses 32 litres of petrol for the drive to Leeds. Julian's car averages 5 km per litre of petrol for the same drive.

  Work out the number of litres Julian's car needs for the drive. **[4]**

Exam tip

When you answer questions involving changing units make sure your final answer is sensible.

CHECKED ANSWERS ☐

Types of triangles and quadrilaterals

Rules

1. A quadrilateral has 4 sides.
2. The 4 angles of a quadrilateral add up to 360°.
3. A triangle with 2 equal angles and 2 equal sides is called isosceles.
4. An equilateral triangle has 3 equal sides and 3 equal angles of 60°.
5. The 3 angles in a triangle add up to 180°.

Worked examples

a Draw a parallelogram, showing its properties.

Answer

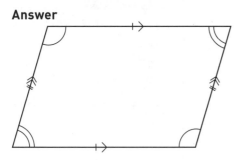

b Which quadrilaterals have
   i   only one pair of equal sides
   ii   two pairs of equal sides
   iii opposite sides parallel and same length.

Answer
   i   trapezium
   ii   square, rhombus, kite
   iii square, rectangle, parallelogram, rhombus

c The diagram shows a rhombus.
   Work out the size of the other angles.

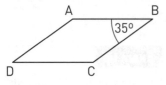

Answer
Angle D = 35° (opposite angles of a rhombus are equal)
Angle A = 180° − 35° = 145° (supplementary angles add
                                              up to 180°)
Angle C = 145° (opposite angles are equal)

d Calculate x

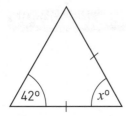

Answer
180° − 42° = 138°
$x° = \frac{1}{2}$ of 138°
            = 69°

> **Exam tip**
>
> You will need to use these markings to show properties of shapes.

> **Exam tip**
>
> You will need to learn all the properties of each type of quadrilateral.

> **Key terms**
>
> Quadrilateral
> Square
> Rectangle
> Parallelogram
> Trapezium
> Kite
> Rhombus
> Diagonal

Exam-style questions

1 PQRS is an isosceles trapezium.
   Work out the values of x and y. **[2]**

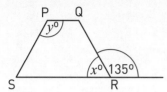

2 ABCD is a quadrilateral.
   AB and CD are parallel and equal. Angle A = Angle C.
   AC = 2BD. What type of quadrilateral is ABCD? **[1]**

3 Calculate y

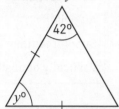

CHECKED ANSWERS

Angles and parallel lines

Rules

❶ Lines that are the same distance apart are called parallel lines.
❷ Corresponding angles are equal.
❸ Alternate angles are equal.
❹ Supplementary angles add up to 180°.
❺ Vertically opposite angles are equal.

Worked examples

a Find the size of each angle marked with a letter and give a reason for each answer.

Answer
$a = 57$ (angles on a straight line add to 180°)

$b = 57$ (alternate angles are equal) ❸

$c = 123$ (vertically opposite angles are equal, supplementary angles add to 180°) ❺ ❹

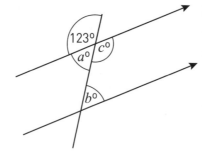

b Find the size of each angle marked with a letter.

Answer
$s = 32$ (32° and s are corresponding angles) ❷

$t = 148$ (s and t are supplementary angles) ❹

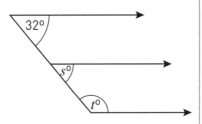

Key terms

Parallel

Corresponding

Alternate

Supplementary

Vertically opposite

Exam-style questions

1 AB is parallel to CD. Write down the values of x and y and give reasons for your answers. **[2]**

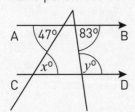

2 PQ and RS are parallel lines. Work out the value of x. **[1]**

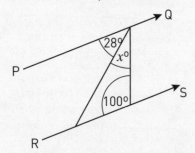

Exam tip

Make sure you give full explanations or reasons if you are asked for them, e.g. don't just write 'alternate angles' write 'alternate angles are equal'.

CHECKED ANSWERS

Angles in a polygon

Rules

1. An interior angle is the angle inside the polygon.
2. The sum of all the interior angles of a n-sided polygon is $180(n - 2)°$
3. The exterior angle is the angle on the outside of a polygon.
4. The sum of all the exterior angles of a polygon is $360°$
5. Exterior angle + interior angle = $180°$
6. In a regular polygon all the interior angles are equal and all the exterior angles are equal.

Worked examples

a Work out the size of the exterior angle of an 18-sided regular polygon.

Answer
Exterior angle = $360° ÷ 18$ (divide $360°$ by the number of sides)
    = $20°$ **4**

b The sum of the interior angles of a regular polygon is $720°$
  i How many sides does the polygon have?
  ii What is the size of the polygon's exterior angle?

Answer
i $180(n - 2) = 720$ **2**
    $n - 2 = \frac{720}{180}$

    $n - 2 = 4$; $n = 6$, so the polygon has 6 sides.

ii exterior angle = $360° ÷ 6 = 60°$ **4** **6**
   exterior angle = $180° - 120° = 60°$ **5**

c Explain why a regular pentagon does not tessellate.

Answer
Sum of interior angles of a pentagon = $180(5 - 2)°$ **2** = $540°$,
        therefore each interior angle = $108°$ **6**
Regular pentagons will not tessellate as 360 cannot be divided by 108 exactly and there will be gaps formed between the shapes.

Key terms

Regular polygon

Interior angle

Exterior angle

Pentagon

Hexagon

Heptagon

Octagon

Decagon

Exam-style questions

1 The diagram shows a regular octagon and a regular pentagon joined along one edge. Calculate the value of x. **[3]**

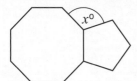

2 Sides AB, BC and CD are 3 sides of a regular n-sided polygon. Work out the values of m and n. You must give reasons for your answers. **[4]**

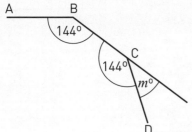

Exam tips

You need to learn the names and properties of all the regular polygons.

Make sure you give reasons if you are asked for them. You will not get some or all of the marks without them.

Finding area and perimeter

Rules

❶ Area is found by multiplying two lengths.
❷ Perimeter is the length around a shape, found by adding its lengths.

Worked examples

a Find the area and perimeter of the rectangle.

Answer

Area = 2 × 5   Perimeter = 5 + 2 + 5 + 2
    = 10 cm²   = 14 cm

an area   a length

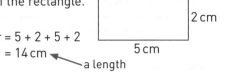

2 cm

5 cm

b Find the area and perimeter of the triangle.

Answer

Area = $\frac{1}{2}$ × length × height

    = $\frac{1}{2}$ × 8 × 6   It is half of a rectangle

    = 24 cm²

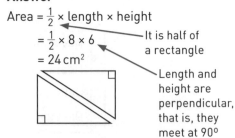

Length and height are perpendicular, that is, they meet at 90°

Perimeter = 6 + 8 + 10
        = 24 cm

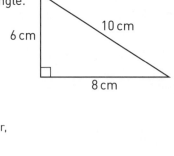

6 cm

10 cm

8 cm

Exam-style questions

1 Find the area and perimeter of the triangle.

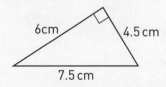

6cm

4.5 cm

7.5 cm

2 Find the area of the parallelogram.

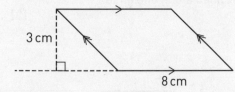

3 cm

8 cm

CHECKED ANSWERS

Circumference and area of circles

Rules

1. The circumference of a circle, C, with a diameter d is $C = \pi d$
2. The circumference of a circle, C, with a radius r is $C = 2\pi r$
3. π has an approximate value of 3.14 or you can use the π button on your calculator.
4. The area of a circle, A, with a radius r is $A = \pi r^2$

Key terms

Circumference

Diameter

Radius

Pi

π

Perimeter

Worked examples

a Find **i** the area **ii** the circumference of a circle with a diameter of 6 cm.

Answer

i $C = \pi d$ **1**
$C = \pi \times 6$
$C = 18.849$ **3** (use the π button on your calculator)
$C = 18.85$ cm to 2 d.p.

ii $A = \pi r^2$ **2**
$A = \pi \times 3 \times 3$
$A = 28.27$
$A = 28.27$ cm to 2 d.p.

b Find the diameter of a circle with a circumference of 25 mm.

Answer
$C = \pi d$ **1**
$C \div \pi = d$
$25 \div \pi = d$ (divide both sides by π)
$7.957 = d$
$d = 7.96$ mm to 3 s.f.

Exam tips

Always write down the unrounded value from your calculator before rounding.

Give answer to 3 s.f. unless you are told otherwise.

c Find the perimeter, p, and area, A, of this shape.

Answer
perimeter of the shape $= \frac{3}{4} \times 2\pi r + 2r$
$p = (0.75 \times 2 \times \pi \times 4) + (2 \times 4)$
$p = 18.849... + 8$
$p = 26.85$ cm to 2 d.p.
$A = \frac{3}{4}\pi r^2$
$A = \frac{3}{4} \times \pi \times 4 \times 4$
$A = 37.7$ cm² to 3 s.f.

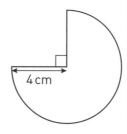

4 cm

Exam tips

Always show all stages of your working.

Do not forget to include units in your answer.

Exam-style questions

1. A wheelbarrow has a front wheel with a diameter of 20 cm.
   A gardener uses the wheelbarrow to move some soil 50 m.
   How many times will the front wheel rotate during the move? **[3]**

2. The diagram shows a window. The window is made from a rectangle and a semicircle.
   The perimeter of the window is 30 m.
   a Calculate the value of x.
   b Calculate the area of the window. **[4]**

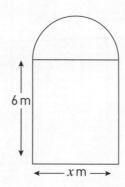

6 m

← x m →

CHECKED ANSWERS ☐

Mixed exam-style questions

1 A recipe requires $\frac{1}{2}$ kg of flour and $\frac{1}{4}$ kg of butter.
What would the approximate imperial equivalents be in pounds? [2]

2 The diagram show the position of two villages.
Redford is on a bearing of 050° from Brownhills.
Karen walks from Brownhills to Redford.
She walks at an average speed of 6 km/h.
She takes 1 h 30 mins to cover the distance.

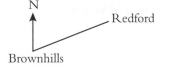

a Work out the distance between Brownhills and Redford. [2]
b Using a scale of 1 cm to 4 km, make an accurate scale drawing showing the position of the two villages. [3]

3 The diagram shows a square attached to four similar regular polygons.

Calculate the number of sides on the polygons. [3]

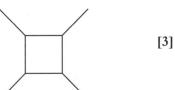

4 The wheel on a bicycle has a diameter of 70 cm.
John cycles 15 km on the bicycle.
a How many revolutions will the wheel make during the journey? [4]
b The journey takes John 1 h 20 min. Calculate John's average speed. [2]

5 Calculate the area of triangle A, rectangle B and triangle C. [4]

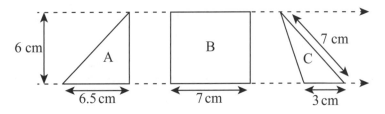

6 The diagram shows a plan of a garden design.
A circular pond with a diameter of 1.5 m is dug in the lawn.
The centre of the pond is in the centre of the lawn.
a Make an accurate scale drawing of the plan using a scale of 2 cm : 1 m. [4]
b Calculate the area of the lawn. [4]

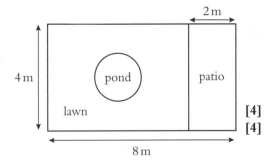

Constructions with a pair of compasses and with a ruler and protractor

REVISED ☐

HIGH

Rules

You can use a pair of compasses and ruler to construct the following:
1. Triangles given their sides
2. Line bisectors
3. Angle bisectors
4. The perpendicular from a point to a line

Worked examples

a Using a ruler, pencil and pair of compasses make an accurate drawing of the triangle ABC.

Answer
Draw the line AB 6.7 cm using a ruler.

Construct a perpendicular at point B. ❹

Construct a perpendicular at point A. ❹

Bisect the angle at A. ❸

Extend the angle bisector from A and the perpendicular at B until they meet at C.

b Using a ruler and pair of compasses, make an accurate drawing of PQR.

Answer
Draw the line PQ 9 cm using a ruler. ❶

Use a pair of compasses 4.5 cm wide at P to draw an arc at Q.

Use a pair of compasses 6 cm wide at R to draw an arc to intersect at Q.

Draw the lines PQ and RQ.

Construct the perpendicular from Q to PR. ❹

c Make an accurate drawing of the triangle and measure the angle marked x.

Answer
Draw a line 8 cm long using a ruler.

Draw the lines from each end of this line.

$x = 100$

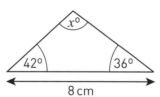

Exam tips

Read the question carefully to check what equipment you may use.

You must show all your construction lines, **do not** rub them out.

Key terms

Arc

Bisector

Perpendicular

Look out for

Make sure all your measurements are very accurate; angles within 2° and lengths within 1 mm. You will lose marks for inaccuracy.

1 a   Using a ruler, pencil and pair of compasses make an accurate
       drawing of the triangle ABC. **[3]**
  The line AX bisects the angle BAC.
  b   Construct the line AX. **[2]**
  c   Measure the distance BX. **[1]**

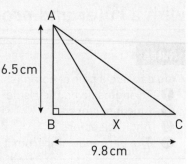

2 Using a ruler, pencil and pair of compasses construct the triangle
   XYZ such that XY = 9 cm, YZ = 6.5 cm and XZ = 7 cm. **[3]**

3 a   Draw a horizontal line PQ 9 cm long and using a pair of
       compasses only, construct the perpendicular bisector of the
       line PQ. Mark the point X where the bisector crosses the line PQ. **[2]**
  b   Mark the point S on the perpendicular bisector so that
       SX = 5.7 cm. **[1]**
  c   Measure the angle SPQ. **[1]**

CHECKED ANSWERS

Translation, reflection and rotation

Rules

1. In a translation the shape does not change orientation. It is picked up and put down elsewhere without turning. The across value is always given first. A positive value means follow the direction of the axis; a negative value means the opposite.

2. In a reflection the shape is mirrored on a line. Think of the mirror line as a fold and the original shape wet ink. The reflection is the shape that this wet ink would now make on the other side of the paper when folded along the mirror line.

3. In a rotation around a point the shape will turn. A point where the turn is centred will be given. Trace the axes and the shape, put your pencil on the centre of the rotation and turn the tracing paper. Look where the rotated shape should be drawn. The axes will not fall on the original axes, unless the rotation was about the origin (0, 0). The traced axes lie vertically and horizontally if the rotation is through 90° or 180°.

3. Turns are given as clockwise or anti-clockwise with an angle in degrees. Clockwise is the direction the hands on a clock turn, and anti-clockwise is the opposite.

Key terms

Anti-clockwise

Clockwise

Co-ordinates

Image

Mirror line

Orientation

Perpendicular

Reflection

Rotation

Translation

Vertex

Worked example

a Triangle A is drawn on a square grid with axes labelled.
   i Translate triangle $A \begin{pmatrix} 2 \\ 3 \end{pmatrix}$, and label the image B.
   ii Reflect triangle A in the line $y = -2$, and label the image C.
   iii Rotate triangle A through 90° anti-clockwise about the point with co-ordinates (0, −1), and label the image D.

Answer

i   Triangle A has been picked up and moved 2 across to the right and 3 up.
ii The line $A = -2$ is drawn by finding −2 labelled on the y-axis and drawing a horizontal line through this point. This gives the mirror line, or line of reflection. The base line of Triangle A is 3 squares away from the line $y = -2$, so the image has to be the same. Remember the fold so the image triangle is upside now.
iii First mark the point (0, −1). Place tracing paper over the square grid. Trace triangle A and the axes. Put the point of your pencil on (0, −1). Now turn the tracing paper 90° anti-clockwise.

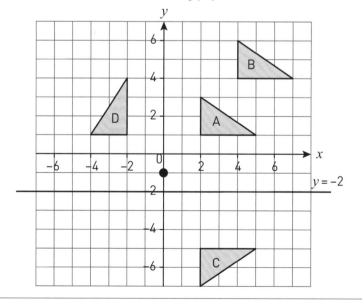

1 Triangle *A* is drawn on a square grid with axes labelled.

   i   Translate triangle $A \begin{pmatrix} -3 \\ -8 \end{pmatrix}$, label the image *B*.

   ii   Reflect triangle *A* in the line $x = -y$, label the image *C*.

   iii Rotate triangle *A* through 90° clockwise about the point with co-ordinates (2, 0), label the image *D*.

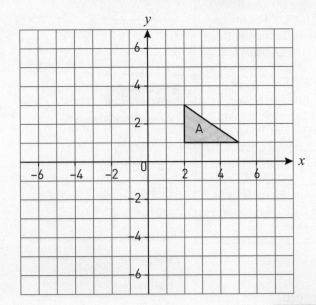

CHECKED ANSWERS

Enlargement

Rules

1. Enlargements are described using a scale factor and centre of enlargement.
2. The scale factor is the amount a shape has been enlarged and can be found by dividing the length of a side of the image by the length of the corresponding side of the object.
3. If the image is smaller than the object the scale factor for the enlargement will be a fraction.

Worked examples

a Plot the points A(2,1), B(4,4), C(5,2) on a set of axes. Join the points to form a triangle. 1 Enlarge the triangle by a scale factor of 2, using (0,0) as the centre of enlargement. Label the points of your image A′ B′ C′.

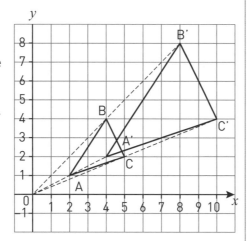

Answer

b Describe fully the transformation that maps shape A on to shape B.

Answer

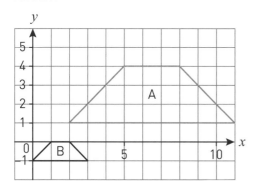

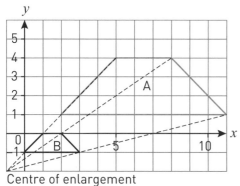

Centre of enlargement

Enlargement, scale factor = $3 \div 9 = \frac{1}{3}$ 2 3

Centre of enlargement = (−1, −2)

Exam tip

When you are asked to perform an enlargement, make sure that corresponding sides are parallel.

Remember

To find the **centre of enlargement,** join the corresponding points of the object and image with straight lines.

The centre of enlargement is where all the lines cross.

Key terms

Object

Image

Scale factor

Centre of enlargement

Map

1 Describe fully the transformation that maps shape
   A on to B. **[2]**

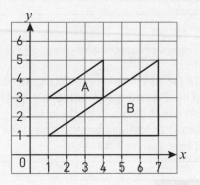

2 Copy the diagram.
   Enlarge the shape Q with a scale factor of $\frac{1}{2}$.
   Centre of enlargement is (0,0).
   Label your answer P.

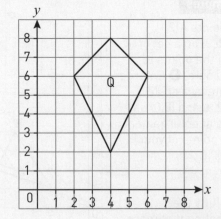

CHECKED ANSWERS

Finding centres of rotation

Rules

1. To describe a rotation fully you need to state the direction, angle and centre of rotation.
2. The centre of rotation is where the perpendicular bisectors of the lines that join the corresponding points of the image and object cross.

Worked examples

a Describe the rotation that maps

  i   A → B
  ii A → C
  iii A → D

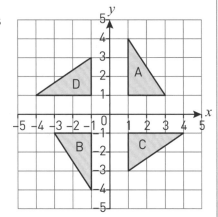

Answer
  i   180° clockwise rotation about (0,0) ❶
  ii 90° clockwise rotation about (0,0) ❶
  iii 90° anti-clockwise rotation about (0,0) ❶

b Image P has been rotated to form image Q.

  i   Find the centre of rotation that maps P to Q.
  ii Describe fully the transformation.

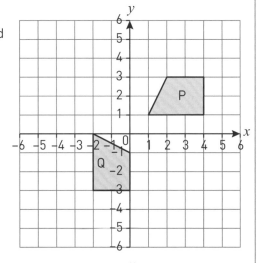

Answer
  i   Centre of rotation is (−1,2) ❶
  ii 180° clockwise rotation about (−1,2) ❷

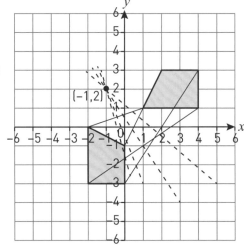

Exam tip

You should leave the lines you used to find the centre of rotation in your answer, as you may get some marks for them if you make a mistake later.

Look out for

Don't forget to write down your answer after finding the centre of rotation.

1 Quadrilateral A has been rotated to position B.
   a Find the centre of rotation that maps quadrilateral A
      to quadrilateral B. **[2]**
   b Describe fully the transformation that maps
      quadrilateral A to quadrilateral B. **[2]**

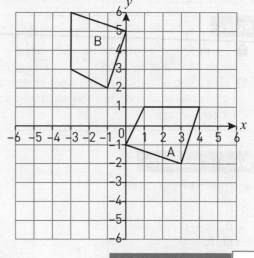

CHECKED ANSWERS

Understanding nets and 2D representation of 3D shapes

HIGH

Rules

1. 2D nets can be folded to make hollow 3D shapes.
2. A plan of a 3D shape is the view from above.
3. An elevation of a 3D shape is the view from the front or side.
4. Isometric paper can be used to make accurate drawings of 3D shapes.

Worked examples

a Make a sketch of the net of this cylinder.

Answer
The net of a cylinder is made from a rectangle and 2 circles. ❶

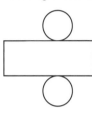

b On squared paper draw the plan and front elevation of the shape shown below. ❹

Answer

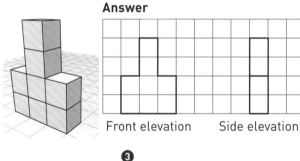

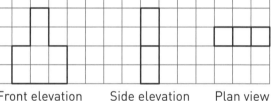

Front elevation   Side elevation   Plan view

❸   ❷

Key terms

Net

2D shape

3D shape

Elevation

Plan view

Isometric

Exam tip

It would be useful to learn the shapes of the nets of a cuboid, pyramid, prism and cylinder.

Look out for

Isometric paper should be the right way up, with vertical lines and no horizontal lines.

Exam-style questions

1 The plan view, front and side elevations for a prism are shown below. On isometric paper draw a 3D representation of the prism. **[3]**

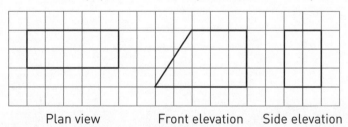

   Plan view   Front elevation   Side elevation

2 The diagram shows a square-based pyramid. Make an accurate drawing of the net of the pyramid. **[5]**

Look out for

If you are asked to make an accurate drawing of a net, you will need to use a ruler and/or pair of compasses.

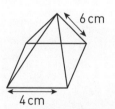

6 cm

4 cm

CHECKED ANSWERS

Geometry and Measures   51

Volume and surface area of cuboids and prisms

REVISED

HIGH

Rules

1. Volume of a cuboid = length × width × height
2. Surface area of a cuboid = 2 × (area of base + area of one side + area of front)
3. Volume of a prism = area of cross-section × height (or length)
4. Surface area of a prism = total area of all the faces
5. Volume of a cylinder = $\pi r^2 h$
6. Surface area of a cylinder = $2\pi rh + 2\pi r^2$

Exam tip

Don't forget to include units in your answers.

Key terms

Volume

Surface area

Prism

Cross-section

Face

Worked example

Work out the volume and surface area of this cuboid.

Answer
Volume = length × width × height ❶

Volume = $8 \times 4 \times 6$

Volume = 192 cm³

Surface area = 2 × (area of base + area of side + area of front) ❷

Surface area = 2 × ((8 × 4) + (4 × 6) + (8 × 6))

Surface area = 2 × (32 + 24 + 48)

Surface area = 208 cm²

Look out for

When you use a calculator to work out problems do not round your answers until the final answer.

Exam-style questions

1 A manufacturer makes stock cubes. The stock cubes are made in 2 cm cubes. She wants to sell the cubes in boxes of 12 and they will be packed with no spaces.
Work out the dimensions of all the possible boxes the manufacturer could choose from. **[3]**

CHECKED ANSWERS

Mixed exam-style questions

1 An isosceles triangle, ABC, has a base, AC, of 8 cm. The perpendicular height of the triangle is 5 cm.

   a Using a ruler and pair of compasses make an accurate drawing of the triangle. **[4]**

   b Measure angle BAC on your drawing. **[1]**

2 Triangle A is drawn on the square grid below.

   a Translate triangle A $\begin{pmatrix} -5 \\ -1 \end{pmatrix}$. Label the image B. **[1]**

   b Rotate triangle A through 90° anticlockwise about the point with coordinates (2, 0). Label the image C. **[2]**

   c Reflect triangle A in the line $y = x$. Label the image D. **[2]**

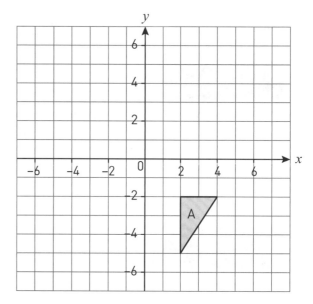

3 Describe fully the transformation that will map shape A onto B. **[2]**

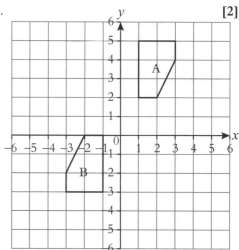

Statistics and Probability: pre-revision check

Check how well you know each topic by answering these questions.
If you get a question wrong, go to the page number in brackets to revise that topic.

1   Charlie recorded the number of goals scored in each of 20 games of football. Here are her results.

Number of goals	1	2	3	4	5
Frequency	5	7	4	2	2

   a Find the median number of goals scored.
   b Work out the mean number of goals scored per game.   (page 56)

2   The table gives information about the number of people living in a village from 1950 to 2010.

Year	1950	1960	1970	1980	1990	2000	2010
Number of people	90	80	100	110	190	240	280

   a Draw a vertical line graph to show the information in the table.
   b Work out the percentage increase in the number of people living in the village between 1980 and 1990.
   c Describe the trend.   (page 58)

3   The table gives some information about the ages of the 54 people on a coach tour.

Age group (in years)	under 20	20 to 39	over 39
Frequency	18	24	12

Lara is going to draw a pie chart to show this information. Work out the angle she should use for the under 20 age group.   (page 59)

4   The table gives information about the age and the trunk radius of each of eight trees.

Age (years)	26	42	50	33	55	58	36	48
Trunk radius (cm)	14	30	42	22	44	52	22	34

   a Draw a scatter diagram to show this information.
   b Describe and interpret the correlation shown in your scatter diagram.
Another tree has an age of 65 years.
   c **i** Find an estimate for the trunk radius of this tree.
      ii How reliable is your estimate? Explain why.   (page 60)

5   Gwen has a hypothesis that more girls use a school library than boys. She also thinks only under-13-year-olds use this library.
      How might you test this hypothesis? You must show an example of how you might find the data you need.   (page 64)

6 Here are some letter tiles.

A	A	A	B	B	C	X

Naomi is going to take at random one of these tiles.
  a What is the probability that Naomi will take the letter A?
  b What is the probability that Naomi will **not** take the letter B?
  (page 65)

7 Mary spins a 4-sided spinner and a 3-sided spinner.
Her score is the difference of the two numbers on the
spinners, as shown in the table.
  a Find the probability that Mary's score is 0.
  b Find the probability that Mary's score is greater than 1.
  (page 66)

		3-sided spinner		
		1	2	3
4-sided spinner	1	0	1	2
	2	1	0	1
	3	2	1	0
	4	3	2	1

8 Jay spins a 3-sided spinner 20 times. Here are her results.

Number	1	2	3
Frequency	9	3	8

Jay is going to spin the spinner one more time.
  a Write down an estimate for the probability that
    the spinner will land on 2.
  b Jay thinks the spinner is biased. Is she right?
    Give a reason for your answer.   (page 67)

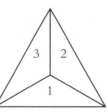

Using frequency tables

Rules

1. The mode is the most common value.
2. The median is the middle value when the data is in order of size.

   The middle value is the $\frac{n+1}{2}$ th value.

3. The range is the difference between the largest and smallest values.
4. To calculate the mean from a frequency table add a $f \times x$ column and a total row to the table.
5. The mean is the sum of all the $f \times x$ values divided by the number of values. Mean $= \frac{\Sigma f \times x}{n}$

Worked example

In a survey 21 people were asked how many loaves of bread they each bought in a week. The table gives information about the results.

Number of loaves (x)	Frequency (f)	$f \times x$
0	4	$0 \times 4 = 0$
1	7	$1 \times 7 = 7$
2	9	$2 \times 9 = 18$
3	1	$3 \times 1 = 3$
Total	21	28

Key term

Frequency

Exam tips

1. Remember that multiplying any number by 0 equals 0
2. If the size of the sample is given in the question (21 in this case) check that the total frequency is equal to it.
3. Give your answers to an appropriate degree of accuracy, generally 3 significant figures or at least 2 decimal places.

Work out
  i   the mode   ii   the median   iii   the mean   iv   the range.

Answer
  i   Mode = 2 ❶ The most common value is the value with the highest frequency. The highest frequency is 9, so the mode is 2

  ii   Median = 1 ❷ The middle value is the $\frac{21+1}{2} = \frac{22}{2} = 11$th value

   There are four 0s, seven 1s, nine 2s, etc., so the 11th value is 1

  iii   Add a $f \times x$ column and a Total row to the table. ❹

   Mean = $\frac{28}{21}$ = 1.33 (2 d.p.) ❺ The sum of all the $f \times x$ values is 28

   The number of values is 20, so the mean = 28 ÷ 21 = 1.33...

  iv   Range = 3 ❸ The largest value is 3, the smallest value is 0, so 3 − 0 = 3

Exam-style questions

1 Mary asks 25 people at a dog show how many dogs they each own. Here are her results.

Number of dogs (x)	Frequency (f)
1	9
2	7
3	5
4	3
5	1

Work out

a the mode **[1]**

b the median **[1]**

c the mean **[3]**

d the range. **[2]**

2 Satbir asked some people how many times they each went to the cinema in the last month. The table shows information about her results.

Number of times (x)	Frequency (f)
0	11
1	15
2	9
3	5

Work out

a the mode **[1]**

b the median **[1]**

c the mean **[3]**

d the range. **[2]**

CHECKED ANSWERS

Statistics and Probability

Rules

1. The scales on the axes do not need to be the same as each other.
2. The scale must be the same along each axis so that the numbers are evenly spaced.
3. If time is involved it goes along the horizontal axis.
4. Label the axes and give the chart a title.

Worked example

The table gives the average highest temperatures recorded in Manchester each month in 2015.

Month	J	F	M	A	M	J	J	A	S	O	N	D
Temp (°C)	6	6	9	12	15	18	20	20	17	14	9	7

Key term

Trend

i   Draw a vertical line chart to show this information.
ii Which two months had the highest average temperatures?
iii Describe the shape of the distribution.

Answer

i

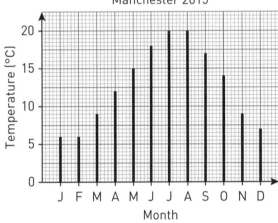

Average highest temperature in Manchester 2015

Exam tip

Be extra careful when interpreting the scale on the vertical axis. Sometimes one square does not represent one unit.

ii The highest average temperatures are given by the tallest lines. The tallest lines are July and August.
iii Average highest temperatures increased from their lowest values in winter, to their highest values in summer, and then decreased again in autumn.

Exam-style questions

1 Fiona sells umbrellas in a shop. The table shows the number of umbrellas she sold each day one week in April.

Day	Mon	Tue	Wed	Thu	Fri	Sat	Sun
Frequency	8	3	4	15	9	6	6

a   Draw a vertical line chart to show this information. **[3]**
b   Fiona sold more umbrellas on Thursday than on any other day. Suggest a reason why. **[1]**
c   The cost of each umbrella is £5.99. How much money did Fiona get for selling umbrellas that week? **[2]**

2 Pam rolled a dice 30 times. Here are her results.
6, 5, 1, 4, 2, 6, 5, 6, 6, 2, 3, 6, 5, 3, 3,
4, 5, 6, 6, 6, 4, 3, 2, 4, 5, 2, 4, 6, 6, 2
a   Draw a vertical line chart to show this information. **[3]**
b   Pam says the dice is biased. Do you agree? Explain why. **[1]**

CHECKED ANSWERS

Pie charts

Rules

1. To draw a pie chart from a frequency table you need to add a 'sector angle' row to the table.
2. The total frequency (n) is the sum of all the frequencies.
3. The angle needed for one item is $\frac{360}{n}$, where n is the total frequency.
4. To calculate the sector angle you multiply the sector frequency (f) by the angle needed for one item: $f \times \frac{360}{n}$
5. To calculate the sector frequency you divide the sector angle (A) by 360 and multiply by the total frequency (n): $\frac{A}{360} \times n$

Worked examples

a The table shows the number of votes Pierre, Carlos, Sasha and Evelyn each got in an election. Draw a pie chart to show this information.

	Pierre	Carlos	Sasha	Evelyn
Number of votes	18	24	33	15
Sector angle	72°	96°	132°	60°

Exam tips

Check your calculations of the sector angles by adding them all up. The total should be 360°.

Label the sectors of your pie chart.

Use a protractor to draw the angles of the pie chart accurately.

Answer
Add a 'sector angle' row to the table ❶. The total frequency (n) = 18 + 24 + 33 + 15 = 90 ❷. Angle needed for one vote = $\frac{360}{90}$ = 4° ❸.

Sector angle for Pierre = 18 × 4 = 72°
Sector angle for Carlos = 24 × 4 = 96°
Sector angle for Sasha = 33 × 4 = 132°
Sector angle for Evelyn = 15 × 4 = 60° ❹

Draw the pie chart.

Key terms

Sector angle

Sector frequency

b The pie chart shows information about the weights of the ingredients needed to make a cake. Kerry uses the information in the pie chart to make a cake. The weight of the cake is 900 grams. How much butter did Kerry use?

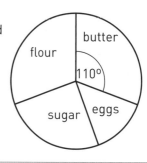

Answer
The sector angle (A) is 110, the total frequency (n) is 900, so
$\frac{110}{360} \times 900 = 275$ grams ❺

Exam-style questions

1 The table shows some information about the drinks sold in a shop one day. Draw a pie chart to show this information. **[4]**

Drink	Tea	Coffee	Juice	Cola
Frequency	12	8	27	25

2 Tony uses the information in the pie chart in Worked example b above to make a cake. He has 385 grams of butter and plenty of the other ingredients. Work out the weight of the largest cake Tony can make. **[2]**

CHECKED ANSWERS

Scatter diagrams and using lines of best fit

Rules

1. An outlier is data that does not fit the pattern of the rest of the data.
2. Positive correlation is when both variables increase together.
3. Negative correlation is when one variable decreases when the other variable increases.
4. When there is correlation between the variables you can draw a line of best fit on the scatter diagram.
5. The line of best fit is the straight line that best represents the data.
6. Use a line of best fit to estimate unknown values.

Worked example

The scatter diagram shows information about the age and the mileage of a sample of 8 cars.

There is an outlier in the data.
  i   Write down the coordinates of the outlier.
  ii Describe any correlation between the age and the mileage of these cars.
  iii Draw a line of best fit on the scatter diagram.
  iv A different car has an age of 4 years. Use your line of best fit to estimate the mileage of this car. Comment on the reliability of your estimate.

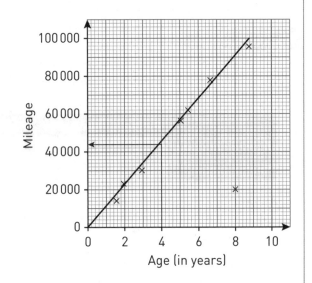

Answer
  i   The data that does not fit the pattern of the other data is (8, 20 000). **1**
  ii The data shows positive correlation, the greater the age of the car the greater the mileage. **2**
  iii See scatter diagram. **4** **5**
  iv The coordinates of the point on the line of best fit that corresponds to an age of 4 years is (4, 44 000). So an estimate for the unknown mileage is 44 000. **6** This estimate uses interpolation. Interpolation is more reliable than extrapolation but, as the sample size is small, the estimate may be unreliable over all.

Exam tips

Show your working for estimates by drawing lines to your line of best fit when reading off values.

Be careful when interpreting scales on the axes of scatter diagrams. One square on the grid does not always represent one unit in the data.

Key terms

Bivariate data

Correlation

Interpolation

Extrapolation

Causation

A scientist recorded the sea temperature at each of eight depths. The table shows information about the results.

Depth (m)	300	150	0	250	400	450	200	100
Sea temperature (°C)	11	15.5	20	13.5	10	8	14	18

a   Draw a scatter graph for this information. **[3]**

b   Describe the correlation. **[1]**

c   i   Estimate the sea temperature at a depth of 350 metres.

   ii   Estimate the depth for a sea temperature of 0 °C. Comment on the reliability of your estimates. **[4]**

CHECKED ANSWERS

Mixed exam-style questions

1 Jose recorded the number of sales he made on his internet shop on each of 40 days. The vertical line chart shows information about his results.

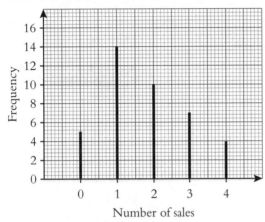

   a Copy and complete the table using the information in the chart.   **[2]**

Number of sales	0	1	2	3	4
Frequency					

   b Find the
     i mode
     ii median
     iii mean of the data.   **[5]**
   c Which average do you think represents the data best?
     Give a reason for your answer.   **[2]**

2 Omar recorded the number of calls to a call centre on each of 5 days. Here are his results.

Day	Mon	Tue	Wed	Thu	Fri
Number of calls	3	5	7	12	18

   a Draw
     i a vertical line chart
     ii a pie chart for this information.   **[6]**
   b Which diagram do you find the most helpful?
     Give a reason for your answer.   **[1]**

3 Emma plays noughts and crosses with her friends.

   She wins $\frac{3}{5}$ of her games,

   loses $\frac{1}{4}$ of her games

   and draws the rest.

   Draw a pie chart to show this information.   **[3]**

4 John recorded the times taken, in minutes, for each of eight students to complete a 250-piece jigsaw puzzle and a 500-piece jigsaw puzzle. His results are given in the table.

250-piece jigsaw (minutes)	35	41	70	71	62	74	45	51
500-piece jigsaw (minutes)	68	70	70	90	86	99	75	78

a Draw a scatter diagram for this information. [3]
b One of the data points may be an outlier. Which data point? Give a reason for your answer. [1]
c Describe and interpret the correlation. [2]

Kyle is another student. It takes him 57 minutes to do the 250-piece jigsaw.

d i Find an estimate for how long it takes him to do the 500-piece jigsaw.
ii Comment on the reliability of your estimate. [3]

Designing questionnaires

Rules

1. Biased questions are leading questions, steering how they might be answered.
2. Vague questions don't allow for useful data to be collated.
3. Options for responses should often be given to avoid vague answers and to help collate the answers.
4. When writing option boxes don't miss any option out, e.g. perhaps an answer of 'don't know' when giving 'yes' and 'no' options.
5. Missing an age or groups of ages out of a questionnaire is a common error.
6. Think about what the survey really wants to find out about as an aim, and write questions to do just this.
7. A questionnaire can be set up to test a hypothesis, which is an idea.

Worked examples

Aled wanted to find out about how often people visited the cinema and the most popular age group of these people.

He stood outside the cinema with the following questionnaire.

1 How old are you?

6 to 10 ☐

12 to 16 ☐

18 to 25 ☐

Over 30 ☐

2 How often do you visit the cinema?

Never ☐

Sometimes☐

Often ☐

Key term

Survey

Bias

Options

Vague

Analyse

Hypothesis

i Explain why Aled's survey could be biased.
ii Write down criticisms of each question in the questionnaire, stating how the questions could be improved.

Answer

i It is carried out outside the cinema, so it is biased towards people who go there.
ii Question 1 overlaps and some ages not covered by the boxes. Consider option boxes with under 6s, 6 to 11-year-olds, 12 to 17-year-olds, 18 to 25-year-olds, 26 to 30-year-olds and 31-year-olds and older
Question 2 is a vague question because it doesn't give any idea of how many times they visit, nor over what period of time. Consider option boxes with number of times visited in a month: 0 to 2, 3 to 5, 6 to 8 and 9 or more.

Exam-style questions

1 Lois has a hypothesis that more boys use the playground in the park than girls. She also thinks only under-8-year-olds use the playground.
How might you test this hypothesis?
You must show an example of how you might find the data you need.

CHECKED ANSWERS ☐

Exam tips

Remember not to be vague. Give options that allow data to be analysed.

Don't miss out any group. Has everyone got an option to tick?

Single event probability

Rules

❶ P(event happening) = $\frac{\text{total number of successful outcomes}}{\text{total number of possible outcomes}}$

❷ P(event not happening) = 1 – P(event happening)

Worked examples

a A letter is going to be picked at random from the word MISSISSIPPI. Find the probability that the letter

i will be an S
ii will not be an S.

Exam tips

Probabilities must be written as fractions, decimals or percentages.

Do not write probabilities as a ratio.

Probabilities written as fractions need not be written in their simplest form.

Answer

i There are 4 Ss, so the total number of successful outcomes is 4
There are 11 letters altogether, so the total number of possible outcomes is 11

P(S) = $\frac{\text{total number of successful outcomes}}{\text{total number of possible outcomes}}$ = $\frac{4}{11}$ ❶

ii P(not S) = 1 – P(S) = $1 - \frac{4}{11} = \frac{7}{11}$ ❷

Key terms

Event

Outcome

Mutually exclusive

b A bag contains some counters. Each counter has a 1, a 2, a 3 or a 4 on it. The table shows information about these counters.

Number on counter	1	2	3	4
Frequency	3	8	7	5

A counter is taken at random from the bag. Work out the probability that the counter has a number greater than 2 on it.

Answer

The total number of successful outcomes = 12, as there are 7 counters with a 3 on them and 5 counters with a 4 on them and 7 + 5 = 12

The total number of possible outcomes = 23, as the total number of counters in the bag = 3 + 8 + 7 + 5 = 23

So P(number greater than 2) = $\frac{\text{total number of successful outcomes}}{\text{total number of possible outcomes}}$ = $\frac{12}{23}$ ❶

Exam-style questions

1 A bag contains 10 counters. 3 of the counters are red, the rest are green. A counter is taken at random from the bag. Write down the probability that the counter
   a will be red **[1]**   b will not be red **[2]**   c will be yellow. **[1]**

2 A weather forecaster says the probability that it will rain tomorrow is 65%. What is the probability that it will not rain tomorrow? **[2]**

3 The sides of a 3-sided spinner are labelled A, B, and C. The probability that the spinner will land on B is twice as likely as it will land on A. The probability it will land on C is twice as likely as it will land on B. Work out the probability that the spinner will land on C. **[3]**

4 A box contains only black counters and white counters. The probability that a counter taken at random from the box will be black is $\frac{5}{12}$. There are 20 black counters in the box. How many white counters are there in the box? **[3]**

CHECKED ANSWERS

Combined events

REVISED

MEDIUM

Rules

❶ Show all the possible outcomes in a list, possibility space or Venn diagram.
❷ Identify all the successful outcomes.
❸ P(event happening) = $\frac{\text{total number of successful outcomes}}{\text{total number of possible outcomes}}$
❹ P(event not happening) = 1 − P(event happening)

Worked examples

a Giles is going to spin a 3-sided spinner numbered 1 to 3 and a 4-sided spinner numbered 1 to 4. Find the probability that the total of the two numbers on the spinners
  i will be 5   **ii** will not be 5.

Answer

i Draw a sample space diagram to show all the possible outcomes. ❶

4-sided spinner

3-sided spinner		**1**	**2**	**3**	**4**
	1	(1, 1)	(1, 2)	(1, 3)	(1, 4)
	2	(2, 1)	(2, 2)	(2, 3)	(2, 4)
	3	(3, 1)	(3, 2)	(3, 3)	(3, 4)

Find all the successful outcomes in the sample space diagram with a total of 5, e.g. (3, 2) ❷
There are 3 successful outcomes, i.e. (3, 2), (2, 3), (1, 4)
There are 12 possible outcomes altogether.
So, P(total 5) = $\frac{\text{total number of successful outcomes}}{\text{total number of possible outcomes}} = \frac{3}{12} = \frac{1}{4}$ ❸

ii P(total not 5) = 1 − P(total 5) ❹
1 − P(total not 5) = $1 − \frac{3}{12} = \frac{9}{12} = \frac{3}{4}$

b The Venn diagram gives information about the number of people in a survey who have watched the films Cinderella (C) and Bambi (B). Some have watched both and some have watched neither. One of these people is picked at random. What is the probability that this person has watched Cinderella or Bambi but not both?

Answer
The Venn diagram shows that 8 people watched only Cinderella, 5 people watched only Bambi, 3 people watched both and 6 people watched neither.
The total number of successful outcomes = 8 + 5 = 13 ❷
The total number of possible outcomes = 8 + 3 + 5 + 6 = 22
So, P(Cinderella or Bambi but not both) = $\frac{\text{total number of successful outcomes}}{\text{total number of possible outcomes}} = \frac{13}{22}$ ❸

Exam-style questions

1 Tim is going to roll a 4-sided spinner numbered 1 to 4 and an ordinary 6-sided dice. Find the probability that the difference of the two numbers he gets will be 2 or less. **[3]**

2 19 people went to Tony's tea shop. Of these 19 people, 12 had a cup of a tea, 15 had a biscuit and 10 had both a cup of tea and a biscuit. One of these people is picked at random. What is the probability they did not have a cup of tea or a biscuit? **[4]**

CHECKED ANSWERS

Estimating probability

REVISED

MEDIUM

Rules

❶ Relative frequency gives an estimate of a probability.

❷ Relative frequency = $\frac{\text{frequency of the event}}{\text{total frequency}}$

❸ The greater the number of trials the greater the reliability of the estimated probability.

Worked examples

a A book shop does a survey to find out if people prefer their books to be hardcopy or digital. The table gives information about the results of the first 237 people surveyed.

	Hardcopy	Digital	Total
Male	59	17	76
Female	108	53	161
Total	167	70	237

Key terms

Event

Bias

Trial

Population

Sample

i Estimate the probability that the next person to be surveyed in the book shop
  a will be male   b will prefer their books to be digital.
ii Do the results show that more people prefer their books to be hardcopy than digital? Give a reason for your answer.

b Omar throws a biased dice 20 times and gets a six 5 times. Jasmine throws the same dice 150 times and gets a six 30 times. Omar and Jasmine each use their results to estimate the probability that the next time they each throw the dice it will land on six. Who has the better estimate, Omar or Jasmine? Explain why.

Answers

a i a 76 males were surveyed, so the frequency of the event is 76.
A total of 237 people were surveyed, so the total frequency is 237.
P(male) = $\frac{\text{frequency of the event}}{\text{total frequency}} = \frac{76}{237}$ ❶ ❷

  b 70 people in the survey prefer digital, so the frequency of the event is 70. A total of 237 people were surveyed, so the total

frequency is 237. P(digital) = $\frac{\text{frequency of the event}}{\text{total frequency}} = \frac{70}{237}$ ❶ ❷

ii No. The results may be biased. Hardcopies of books are sold in book shops, so there may be more people in the book shop who prefer their books to be hardcopies.

b Jasmine has the better estimate as she threw the dice more times than Omar. ❸

Exam tips

Probabilities must be written as fractions, decimals or percentages.

Probabilities written as fractions need not be written in their simplest form.

Exam-style questions

1 Hilary spins a biased coin 20 times. Here are her results.
H, T, H, H, T, H, H, T, H, T, H, T, H, H, H, H, H, H, T, H
  a Find an estimate for the probability that the next time she spins the coin it will land on Heads. **[2]**
  b Hilary is going to spin the coin 300 times. Work out an estimate for the number of times the coin will land on Tails. **[2]**

2 In a survey of tourists a sample of 60 people were asked to choose to go on a Tower of London tour or on a Westminster Abbey tour. 29 males were sampled of whom 15 chose to go on the Tower of London tour. 17 females chose to go on the Westminster Abbey tour. Find an estimate for the probability that the next person to be surveyed will choose to go on the Tower of London tour. **[4]**

CHECKED ANSWERS

Mixed exam-style questions

1 Haley is going to pick at random a letter from the word STATISTICS.
   Write down the probability that the letter will be S.   [1]

2 The probability that Mark will get a blue paper hat in a Christmas cracker is 0.85.
   Work out the probability he will not get a blue paper hat in the Christmas cracker.   [2]

3 An insurance company received a total of 3467 claims last year. Of these 2125 were for household
   damage. This year the insurance company expects to receive a total of 5000 claims.
   Estimate the number of claims for household damage this year.   [2]

4 Cheri spins a fair 5-sided spinner, numbered 1, 2, 3, 4 and 5, and rolls an ordinary dice.
   What is the probability that Cheri will get
   a a 3 on the spinner and a 3 on the dice   [2]
   b a 3 on the spinner or a 3 on the dice?   [2]

5 Bag A contains 3 red counters and 2 green counters. Bag B contains 4 red counters and 5 green
   counters. Bag C contains 1 red counter and 6 green counters. Hamish is going to take at random a
   counter from bag A. If he takes a red counter he will take a counter at random from bag B. If he takes
   a green counter he will take a counter at random from bag C.
   Work out the probability that both counters will be the same colour.   [4]

6 Nima rolls 5 ordinary dice. Work out the probability that he will get exactly three 6s.   [3]

7 Rhod wanted to find out about how often people use the train to travel to work and the most popular
   age group of these people.
   He stood outside the railway station between 08:15 and 09:00 with the following questionnaire.
   He asked people leaving the station to answer his questions.

1 How old are you?	
Under 16	☐
16 to 30	☐
30 to 40	☐
Over 40	☐
2 How often do you use the train to travel to work?	
Never	☐
Sometimes	☐
Often	☐

   a Explain why Rhod's survey could be biased.   [1]

   b Write down criticisms of each question in the questionnaire, stating how the questions could be
          improved.   [4]

The language used in mathematics examinations

- **You must show your working...** you will lose marks if working is not shown.

- **Estimate...** often means round numbers to 1 s.f.

- **Calculate...** some working out is needed; so show it!

- **Work out/find...** a written or mental calculation is needed.

- **Write down...** written working out is not usually required.

- **Give an exact value of...** no rounding or approximations:
  - on a calculator paper, write down all the numbers on your calculator.
  - on a non-calculator paper, give your answer in terms of π.

- **Give your answer to an appropriate degree of accuracy...** if the numbers in the question are given to 2 d.p. give your answer to 2 d.p.

- **Give your answer in its simplest form...** usually cancelling of a fraction or a ratio is required.

- **Simplify...** collect like terms together in an algebraic expression.

- **Solve...** usually means find the value of x in an equation.

- **Expand...** multiply out brackets.

- **Construct, using ruler and compasses...** the ruler is to be used as a straight edge and compasses must be used to draw arcs. You **must** show all your construction lines.

- **Measure...** use a ruler or a protractor to accurately measure lengths or angles.

- **Draw an accurate diagram...** use a ruler and protractor – lengths must be exact, angles must be accurate.

- **Make y the subject of the formula...** rearrange the formula to get y on its own on one side e.g. $y = \frac{2x - 3}{4}$

- **Sketch...** an accurate drawing is not required – freehand drawing will be accepted.

- **Diagram NOT accurately drawn...** don't measure angles or sides – you must calculate them if you are asked for them.

- **Give reasons for your answer... OR explain why...** worded explanations are required referring to the theory used.

- **Use your/the graph...** read the values from your graph and use them.

- **Describe fully...** usually transformations:
  - Translation
  - Reflection in a line
  - Rotation through an angle about a point
  - Enlargement by a scale factor about a point

- **Give a reason for your answer...** usually in angle questions, a written reason is required e.g. 'angles in a triangle add up to 180°' or 'corresponding angles', etc.

- **You must explain your answer...** a worded explanation is required along with the answer.

- **Show how you got your answer...** show all your working. Words may also be needed.

- **Describe...** answer the question using words.

- **Write down any assumption you make...** describe any things you have assumed are true when giving your answer.

- **Show...** usually requires you to use algebra or reasons to show something is true.

Exam technique and formulae that will be given

- Be prepared and know what to expect.

- Don't just learn key points.

- Work through past papers. Start from the back and work towards the easier questions. Your teacher will be able to help you.

- Practice is the key, it won't just happen.

- Read the question thoroughly.

- Cross out answers if you change them, only give **one** answer.

- Underline the key facts in the question.

- Estimate the answer.

- Is the answer right/realistic?

- Have the right equipment:
  - Calculator
  - Pens
  - Pencils
  - Ruler, compass, protractor
  - Eraser
  - Tracing paper
  - Spares

- Never give two different answers to a question.

- Never just give just an answer if there is more than 1 mark.

- Never measure diagrams; most diagrams are not drawn accurately.

- Never just give the rounded answer; always show the full answer in the working space.

- Read each question carefully.

- Show stages in your working.

- Check your answer has the units.

- Work steadily through the paper.

- Skip questions you cannot do and then go back to them if time allows.

- Use marks as a guide for time: 1 mark = 1 min

- Present clear answers at the bottom of the space provided.

- Go back to questions you did not do.

- Read the information below the diagram – this is accurate.

- Use mnemonics to help remember formula you will need, for example:
  - For the order of operations, BIDMAS: Brackets, Indices, Division, Multiply, Add, Subtract
  - Formula triangles for the relationship between three parameters e.g. speed, distance and time

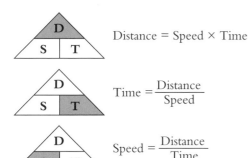

$$\text{Distance} = \text{Speed} \times \text{Time}$$

$$\text{Time} = \frac{\text{Distance}}{\text{Speed}}$$

$$\text{Speed} = \frac{\text{Distance}}{\text{Time}}$$

Common areas where students make mistakes

Here are some topics that students frequently make errors in during their exam.

Number

> **Rules**
> 1. A factor is a number that divides into another number, e.g. 2 is a factor of 6.
> 2. A multiple is a member of the multiplication table of that number, e.g. 6 is a multiple of 2.
> 3. A prime number is one that can only be divided by 1 and itself, e.g. 2, 3, 5, 7, 11 ...

	Question	Working	Answer
Adding	$\frac{2}{3}+\frac{1}{4}$	Write equivalent fractions: $\frac{2}{3}=\frac{4}{6}=\frac{6}{9}=\frac{8}{12}$ and $\frac{1}{4}=\frac{2}{8}=\frac{3}{12}$ 12 is the LCM of 3 and 4 so write the fractions in 12ths: $\frac{8}{12}+\frac{3}{12}=\frac{8+3}{12}=\frac{11}{12}$	$\frac{11}{12}$
Subtracting	$\frac{2}{3}-\frac{1}{4}$	You use the same method as adding but just take away, so we get: $\frac{8}{12}-\frac{3}{12}=\frac{8-3}{12}=\frac{5}{12}$	$\frac{5}{12}$
Multiplying	$\frac{2}{5}\times\frac{3}{8}$	Multiply the tops together and then the bottoms of the fractions: $\frac{2\times3}{5\times8}=\frac{6}{40}$ then cancel by 2	$\frac{3}{20}$
Dividing	$\frac{5}{12}\div\frac{2}{3}$	Write the first fraction down and turn the second fraction upside down and multiply: $\frac{5}{12}\times\frac{3}{2}=\frac{15}{24}$ then cancel by 3	$\frac{5}{8}$

	Question	Working	Answer
Changing a fraction to a decimal	Write $\frac{3}{8}$ as a decimal.	Divide the top number by the bottom number so divide 3 by 8: $\begin{array}{r} 0.375 \\ 8\overline{)3.0^30^60^40} \end{array}$	0.375

	Question	Working	Answer
Finding a fraction of an amount	Find $\frac{3}{5}$ of £4.80	This can be written as $\frac{3}{5} \times £4.80$ There is a simple rule for this calculation, which is 'Divide by the bottom and Times by the top' You can sing it to the 'Wheels on the bus' song to help you remember it. £4.80 ÷ 5 = 0.96 then 0.96 × 3 = £2.88	£2.88
Finding a percentage of an amount	Work out 60% of £4.80	This can be written as $\frac{60}{100} \times £4.80$ You can use the same rule so you divide by 100 and times by 60: £4.80 ÷ 100 × 60 = £2.88	£2.88

	Question	Working	Answer
Estimating	Estimate $\frac{76.15 \times 0.49}{19.04}$	Write each number to one significant figure so that: 76.15 becomes 80 0.49 becomes 0.5 19.04 becomes 20 Remember that the size of the estimate needs to be similar to the original number. So 80 × 0.5 = 40 and 40 ÷ 20 = 2	2

	Question	Working	Answer
Using a calculator	Work out $\frac{76.15 + 5.62^2}{19.04}$	You either need to enter the whole calculation into your calculator using the fraction button or work out the top first then divide the answer by the bottom. $76.15 + 5.62^2 = 107.7344$ 107.73 ÷ 19.04 = 5.658 319 327	5.658 319 327

Algebra

Rules

$2y + y = 3y$ $2y - y = y$ $2y \times y = 2y^2$ $2y \div y = 2$

$y^m \times y^n = y^{m+n}$ $y^m \div y^n = y^{m-n}$ $(y^m)^n = y^{mn}$

	Question	Working	Answer
Collecting like terms	Simplify		
	a $2ab + 3ab - ab$	$= 2ab + 3ab - 1ab$	$4ab$
	b $3y^2 - y^2$	$= 3y^2 - 1y^2$	$2y^2$

	Question	Working	Answer
Multiplying out the brackets	Expand		
	a $3(4p + 5)$	$= 3 \times 4p + 3 \times 5$	$12p + 15$
	b $7p - 4(p - q)$	$= 7p - 4 \times p - 4 \times -q = 7p - 4p + 4q$	$3p + 4q$
	c $(y + 3)(y - 4)$	$= y \times y + y \times - 4 + 3 \times y + 3 \times - 4$	
		$= y^2 - 4y + 3y - 12$	$y^2 - y - 12$

	Question	Working	Answer
Solving equations	Solve		
	a $3t - 2 = 4$	$3t = 4 + 2$ so $3t = 6$	$t = 2$
	b $3f + 4 = 5f - 3$	$4 + 3 = 5f - 3f$ so $7 = 2f$ or $2f = 7$	$f = 3.5$
	c $5(x + 2) = 3$	$5x + 10 = 3$ so $5x = 3 - 10$ or $5x = -7$	$x = -1.4$
	d $y^2 - 3y - 10 = 0$	$(y + 2)(y - 5) = 0$ so $y + 2 = 0$ or $y - 5 = 0$	$y = -2$ or $y = 5$

Geometry and Measures

Rules

The **perimeter** of a shape is the distance around its edge. You **add** all the side lengths together.

The **area** of a shape is the amount of flat surface it has. You **multiply** two lengths.

The **volume** of a shape is the amount of space it has. You **multiply** three lengths.

Alternate angles are in the shape of a letter **Z**.

Corresponding angles are in the shape of a letter **F**.

Allied angles or co-interior angles are in the shape of a letter **C**.

	Question	Working	Answer
Perimeter of a shape	Find the perimeter of this shape. 3 cm ▭ 5 cm	For a rectangle you need to add the lengths of the four sides. $3 + 5 + 3 + 5 = 16$	16 cm

	Question	Working	Answer
Area of a shape	Find the area of this shape. 3 cm ◺ 6 cm	For this right-angled triangle you need to use the formula: Area $= \frac{1}{2}$ base $\times$ vertical height So the area $= \frac{1}{2} \times 6 \times 3 = 9$ You have multiplied two lengths.	9 cm²

	Question	Working and answer
Angles between parallel lines	Find the missing angles in this diagram. Give reasons for your answer.	$a = 50°$ (Alternate angles are equal) $b = 130°$ (Allied angles add to 180° (supplementary)) $c = 50°$ (Corresponding angles are equal)

	Question	Working and answer
Finding missing angles and giving reasons	ABC is an isosceles triangle. BCD is a straight line. Find, giving reasons, angle ACD.	Angle ABC = (180 − 50) ÷ 2 = 65° (The three angles of a triangle add to 180°) Angle ACB = Angle ABC = 65° (Base angles of an isosceles triangle are equal) Angle ACD = 180 − 65 = 115° (Sum of the angles on a straight line = 180°)

Statistics and Probability

	Question	Working	Answer
Pie chart	Draw a pie chart from this information. **Favourite colour / f** Red 7 Blue 4 Green 2 Yellow 3 Black 4	As pie charts are based on a circle then we need to divide the number of degrees in a whole turn (360°) by the total frequency which is 20. So 360° ÷ 20 = 18° The angle for each colour is then calculated by multiplying its frequency by 18°.	Red = 7 × 18° = 126° Blue = 4 × 18° = 72° Green = 2 × 18° = 36° Yellow = 3 × 18° = 54° Black = 4 × 18° = 72° Then draw the circular pie chart.

One week to go

You need to know these formulae and essential techniques.

Number

Topic	Formula		When to use it
Negative numbers	+ + = +	– – = +	Two signs next to each other
	+ – = –	– + = –	
	+ × + = +	– × – = +	Multiplying integers
	+ × – = –	– × + = –	
	+ ÷ + = +	– ÷ – = +	Dividing integers
	+ ÷ – = –	– ÷ + = –	
Order of operations	BIDMAS		If you have to carry out a calculation. You use the order Brackets, Indices, Division, Multiplication, Addition and Subtraction.
Percentages	20% of $50 = \frac{20}{100} \times 50$		To find the percentage of an amount e.g. 20% of 50.
Simple interest	SI for 5 years at 3% on £150 $\frac{3}{100} \times 150 \times 5$		To find the **simple interest** you find the interest for one year and multiply by the number of years.
Standard form	$2.5 \times 10^3 = 2500$ $2.5 \times 10^{-3} = 0.0025$		A number in standard form is (a number between 1 and 10) × (a power of 10)
Approximating	Decimal places		You round to a number of decimal places by looking at the next decimal place and rounding up or down.

Geometry and Measures

Topic	Formula	When to use it
Parallel sides		Parallel lines are shown with arrows.
Equal sides		Equal lines are shown with short lines.
Perimeter	Add lengths of all sides.	To find the perimeter of any 2D shape

Areas of 2D shapes	Area = $l \times w$	Area of a rectangle is length × width w l
	Area = $\frac{1}{2}b \times h$	Area of a triangle is $\frac{1}{2}$base × vertical height h b
	Area = $b \times h$	Area of a parallelogram is base × vertical height h b
	Area = $\frac{1}{2}(a + b) \times h$	Area of a trapezium is $\frac{1}{2}$ the sum of the parallel sides × the vertical height a h b
Circumference and area of a circle	$C = \pi \times D$ or $C = \pi \times 2r$	Circumference or the perimeter of a circle is: pi × diameter **or** pi × double the radius r D
	$A = \pi \times r^2$	Area of a circle is pi × radius squared
Volumes of 3D shapes	$V = l \times w \times h$	Volume of a cuboid is: Length × width × height

Answers

Number

Pre-revision check (page 1)

1 a 8.5   b 18   c 1.4
2 a 25.392   b 56.7
3 a 26.1   b 26.38
4 a 0.018   b 124.5   c 254 900   d 48.7
5 a 5.72×10^{-3} b 31 840
6 a 16.4   b 16.35   c 16.355
7 a 3000   b 0.003
8 a $\frac{5}{18}$ b $\frac{3}{55}$ c 9   d $\frac{1}{2}$
9 a $\frac{11}{60}$ b $3\frac{5}{12}$ c $18\frac{4}{15}$ d 10   e $1\frac{1}{5}$
10 a i 55%   ii 190%   b i $\frac{3}{50}$ ii 0.06
11 a £13   b 374 m
12 a 8%   b 25%
13 a 105 m, 175 m   b $\frac{5}{9}$
14 1976 g or 1.976 kg
15 a 3^5 b 3^5 c 3^{18}
16 $2 \times 2 \times 3 \times 3 \times 5 \times 7$

BIDMAS (page 2)

1 a 4   b $3 + (9 - 5) \times 2 = 11$
2 $7 - 10 \div (3 + 2) = 5$ 3 0.9435...

Multiplying decimals and negative numbers (page 3)

1 614.6
2 Shop A (£64.32) is cheaper than Shop B (£64.80)
3 £10.48
4 2

Dividing decimals and negative numbers (page 4)

1 £7.92
2 6 (6.65)
3 £253.60
4 −3

Using the number system effectively (page 5)

1 a 0.539   b 4580
2 a 4165   b 11.9   c 0.833

Understanding standard form (page 6)

1 a 7.2×10^{-2} b 2.389×10^5
2 a 9 140 000   b 0.000 518   3 17×10^{-2}

Rounding to decimal places and approximating (page 7)

1 11.44 cm²
2 2000

Multiplying and dividing fractions (page 8)

1 a $\frac{1}{48}$ b $\frac{20}{9}$
2 a $\frac{1}{6}$ b $\frac{2}{15}$

Adding and subtracting fractions and working with mixed numbers (page 9)

1 a $\frac{23}{24}$ b $2\frac{5}{6}$
2 a $\frac{5}{8}$ b $\frac{9}{40}$
3 $11\frac{1}{3}$ m²

Converting fractions and decimals to and from percentages (page 10)

1 20%   0.202   0.21 $\frac{2}{9}$ $\frac{1}{4}$
2 0.031 818 18...   3 21.25%

Calculating percentages and applying percentage increases and decreases to amounts (page 11)

1 12.75 cm
2 £405 in the auction, so could have £5 more
3 108 before and 104 after, so fewer

Finding the percentage change from one amount to another (page 12)

1 2.5%   2 3.75% loss
3 Nazia (15%) more than Debra (14.5%)

Mixed exam-style questions (page 13)

1 $5 \times (2 + 3) - 7 = 18$. $5 \times 2 + (3 - 7) = 6$
2 No, Naomi = £446.95 and Izmail = £441
3 a tube = 1.2 p per sweet, box = 1.25 p per sweet
   b 1 box and 2 tubes
4 a 1.6107   b 158.34
5 a $0.5 \times 7.6 \times 4.0 = 15.2$ cm²
   b less since both dimensions are less
6 £7200

Sharing in a given ratio (page 14)

1 £60, £48, £24
2 6384   3 Small packet since £4.30 ÷ 3 = £1.43 per share, £6.30 ÷ 4 = £1.58 per share and £7.50 ÷ 5 = £1.50 per share

Working with proportional quantities (page 15)

1 £9.75   2 £1.35   3 88

Index notation (page 16)

1 a 10^4 b 32768   2 2^{15}

3 $10^4 \times 10^5 = 10^9$, $10 \times 10^2 = 10^3$, $\dfrac{10^{20}}{10^2} = 10^{18}$, $10^{10} = 10\,000\,000\,000$

Prime factorisation (page 17)

1 a $2 \times 2 \times 2 \times 2 \times 2 \times 3$
   b i 24   ii 480

2 Saturday 8 am

Mixed exam-style questions (page 18)

1 a $\dfrac{5}{8}$ b $\dfrac{1}{9}$

2 $3\dfrac{13}{63}$ hours

Algebra

Pre-revision check (page 19)

1 8

2 a $a = 2$ b $b = 15$ c $c = \dfrac{2}{5}$ d $e = -6$

3 a i $10a + 15$ ii $3h^2 - 6h$ iii $12x^2 - 6xy$
   b i $6(v + 2)$ ii $3p(2p - 3)$
      iii $5e(e + 2f)$ iv $4xy(2x - 3y)$

4 a $x = 3$ b $p = -0.75$ c $y = -1\dfrac{1}{11}$

5 a $g = 3$ b $h = -\dfrac{2}{7}$ c $k = -7$

6 a $6n - 2$ b 298
   c If $6n - 2 = 900$ then $6n = 902$ and $n = 150.3$.
      The term numbers must be whole numbers so
      900 is not a member of the sequence.

7 a

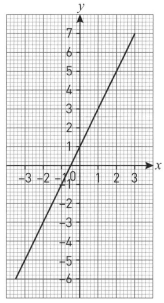

   b $x = 2.5$

8 08:00 and 12:48

Working with formulae (page 20)

1 a 10   b 90
2 a 2 hours or 120 minutes   b 4.5 kg

3 Rachel since Rachel's = £35 880
   and Peter's = £34 874

4 a 1363
   b 23.5%

5 $\dfrac{6}{35}$

6 a Bill, since Bill = 1.25 m³ Sandra = 1.23 m³
   b Less needed, since $\dfrac{2}{19} < \dfrac{1}{8}$

7 Medium, since for small bottle 1 pint = £1.24,
   for medium bottle 1 pint = £1.22, for large
   bottle 1 pint = £1.23

Setting up and solving simple equations (page 21)

1 a $a = 10$ b $b = 15$ c $c = -\dfrac{2}{3}$
   d $d = 10$ e $e = 1.5$ 2 76 cm²

Using brackets (page 22)

1 Amy is 8, Beth is 11 and Cath is 22
2 $x = 8.5$; 36 cm

More complex equations and solving equations with the unknown on both sides (page 23)

1 104 cm²
2 Equating two angles: $2x + 30 = 5x - 15$
   Solving: $45 = 3x$; $x = 15$
   OR total of angles is $10x + 30 = 180$
   Solving: $10x = 150$; $x = 15$
   Substituting $x = 15$ into each expression gives
   60° for each angle. Therefore triangle ABC is
   equilateral. I have assumed that each angle of an
   equilateral triangle is 60°.

Solving equations with brackets (page 24)

1 $n = 6$ 2 2250 cm²

Linear sequences (page 25)

1 a 27   b $4n + 3$
   c $4n + 3 = 163$ so $4n = 160$ and $n = 40$ so 163 is
      a term in the sequence
2 nth term is $59 - 4n$ so $59 - 4n < 0$ so $59 < 4n$
   therefore $n > 14.75$ which makes $n = 15$

Mixed exam-style questions (page 26)

1 a 37°C b −40°C
2 10 20 am
3 $\dfrac{40x + 60}{4} = 10x + 15$, $\dfrac{40x + 60}{5} = 8x + 12$;
   $10x + 15 - (8x + 12) = 2x + 3$
4 17
5 a $n + n + 1 + n + 2 + 10 + n + 1 + 20 + n + 1 =$
      $5n + 35 = 5(n + 7)$
   b If $5n + 35 = 130$; then $n = 19$. n cannot equal
      19 because it will overlap the grid.

Plotting graphs of linear functions (page 28)

1 a

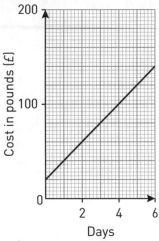

b 6 days

2 a

x	−3	−2	−1	0	1	2	3
y	−7	−5	−3	−1	1	3	5

b

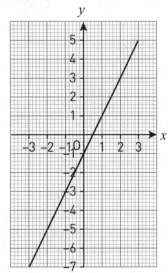

c $x = 2.5$

Real-life graphs (page 29)

a 08 00   **b** 15 minutes   **c** 30 minutes
d 10 15   **e** 9 miles   **f** 1 hour
g 45 minutes   **h** 12 miles per hour

Mixed exam-style questions (page 31)

1 a £80   **b** £15
  c Draw line on graph from $(0, 0)$ to $(5, 125)$ or compare cost for each day
     Up to 2 days cheaper with Car Co, 3 days or more is cheaper with Cars 4 U

2

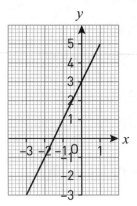

Geometry and Measures

Pre-revision check (page 31)

1 100 miles
2 a   **b**   055°

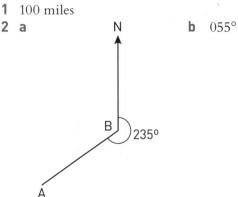

3 a 17 m   **b** 14 cm   **4** 86.4 m
5 a Trapezium   **b** Square, rhombus
  c Rectangle, parallelogram, rhombus
6 $a = 67°$, $b = 67°$
7 exterior angle = 30°, interior angle = 150°
8 Area: 48 cm²
    Perimeter: 30 cm
9 area = 78.54 cm², circumference = 31.42 cm
10

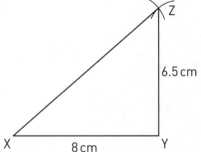

11

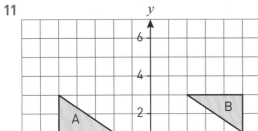

12 90° anti-clockwise rotation about (3,−1)

13 a

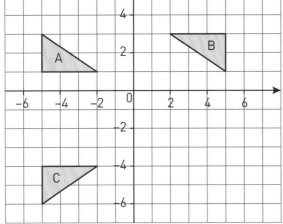

Net   Plan   Side

b

Net   Plan   Side

14 40 cm³, 76 cm²
15 254 mm³, 226 mm²
16

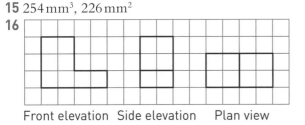

Front elevation   Side elevation   Plan view

Converting approximately between metric and imperial units (page 33)

1 480 pints
2 552 km

Bearings and scale drawings (page 34)

1 a Accurate drawing
  b 228°
2 a 2 km   **b** 26.8 cm   **3** 260°

Compound units (page 35)

1 918 km/hr
2 400 secs or 6 mins 40 sec   **3** 64 litres

Types of triangles and quadrilaterals (page 37)

1 $x = 45°$, $y = 135°$
2 ABCD is a rhombus or a parallelogram
3 96°

Angles and parallel lines (page 38)

1 $x = 47°$ ($x°$ and 47° are alternate angles are
  equal), $y = 97°$ ($y°$ and 83° are supplementary
  angles, add up to 180°)
2 $x = 52°$

Angles in a polygon (page 39)

1 117°
2 $m = 36°$, sum of interior angle and exterior
  angle = 180°
  sum of exterior angles of a polygon = 360°
  therefore number of exterior angles = 360 ÷ 36,
  therefore $n = 10$

Finding area and perimeter (page 40)

1 perimeter: 18 cm, area: 27 cm²
2 24 cm²

Circumference and area of circles (page 41)

1 79.6 rotations   **2 a** 7.00 m   **b** 61.3 m²

Mixed exam-style questions (page 42)

1 $\frac{1}{2}$ kg ≈ 1.1 pounds and $\frac{1}{4}$ kg ≈ 0.55 pounds
2 a 9 km   **b**

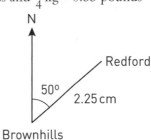

3 8 sides
4 a 6820 revs to 3 s.f.
  b 11.25 km/hr
5 Area of A: 19.5 cm²
  Area of B: 42 cm²
  Area of C: 9 cm²
6 a Scale drawing
  b 22.2 m² to 3 s.f.

Constructions with a pair of compasses and with a ruler and protractor (page 44)

1 a Accurately constructed right angled triangle
  b Accurately constructed angle bisector at A
  c BX = 3.5 cm
2 Accurate drawing of triangle XYZ such that
  XY = 9 cm, YZ = 6.5 cm and XZ = 7 cm.
3 a Perpendicular bisector of line PQ,
  (PX = XQ = 4.5 cm)
  b SX drawn = 5.7 cm   **c** SPQ = 52°

Translation, reflection and rotation (page 46)

1 $x = -y$

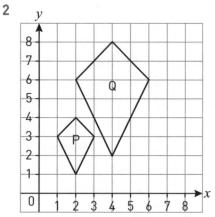

Enlargement (page 48)

1 Enlargement, sf 2, centre of enlargement (1, 5)

2

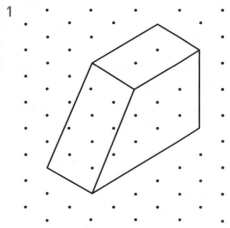

Finding centres of rotation (page 50)

1 a (−2,0)

b 90° anti-clockwise rotation about (−2, 0)

Understanding nets and 2D representation of 3D shapes (page 51)

1

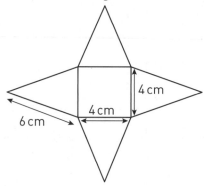

2 Accurate drawing of net shown below

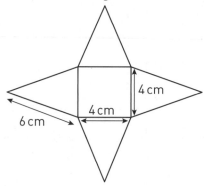

Volume and surface area of cuboids and prisms (page 52)

1 24 cm × 2 cm × 2 cm. 4 cm × 4 cm × 6 cm. 4 cm × 12 cm × 2 cm. 2 cm × 8 cm × 6 cm

Mixed exam-style questions (page 53)

1 a Accurate drawing   b 51°

2

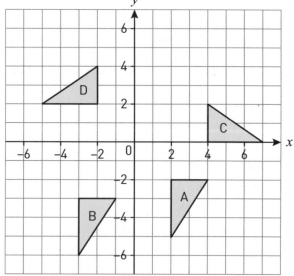

3 rotation 180° about (0, 1)

Statistics and Probability

Pre-revision check (page 54)

1 a 2   b 2.45

2 a

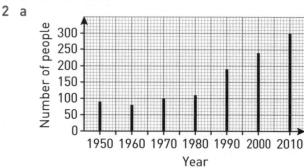

b 72.7%

c increasing

3 120°

4 a

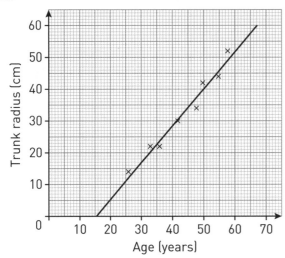

b Positive correlation, since the older the tree the greater the trunk radius

c i 55 – 65 cm

ii Not reliable - extrapolation or Could be reliable since it is not that far away from the original data values

5 For example, ask students in the library their gender and age.
A questionnaire might ask:
What gender?   Boy ☐ Girl ☐
How old are you?
           Under 11   ☐
      11 to 12 years old   ☐
      13 to 14 years old   ☐
      15 to 16 years old   ☐
           Over 16   ☐

6 a $\frac{3}{7}$ **b** $\frac{5}{7}$

7 a $\frac{3}{12}$ **b** $\frac{4}{12}$

8 a $\frac{3}{20}$

b Either "yes, because the frequencies should be roughly equal" or "no, not enough spins to conclude whether this is true"

Using frequency tables (page 57)

1 a 1   **b** 2   **c** 2.2   **d** 4

2 a 1

b 1

c 1.2

d 3

Vertical line charts (page 58)

1 a

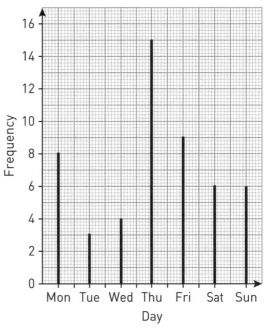

b It rained   **c** £305.49

2 a

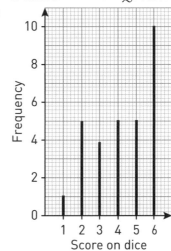

b Yes, as significantly more 6s than other numbers OR Can't tell as more throws needed

Pie charts (page 59)

1 Tea: 60°, coffee: 40°, juice: 135°, cola: 125°

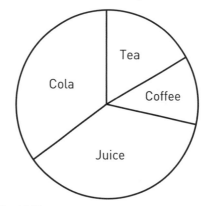

2 1260 grams

Scatter diagrams and using lines of best fit (page 60)

a

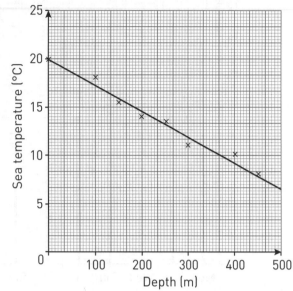

b Negative correlation; as the depth increase the sea temperature decreases

c i 10.5° C. Interpolation, so reliable (although small amount of data may make this unreliable)

  ii 700 metres. Extrapolation, so unreliable

Mixed exam-style questions (page 62)

1 a 5, 14, 10, 7, 4

  b i 1 ii 2 iii 1.775

  c e.g. mean, as it includes all the data

2 a i

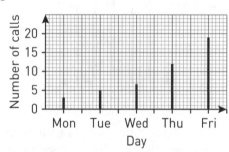

  ii Mon: 24°, Tue: 40°, Wed: 56°, Thu: 96°, Fri: 144°

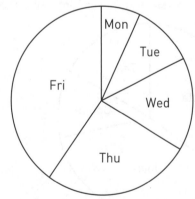

  b Line chart, because it shows frequencies OR Pie chart, because it shows proportions

3 Pie chart: angles Wins 216°, Loses 90°, Draws 54°; each sector labelled appropriately Wins, Loses or Draws.

4 a

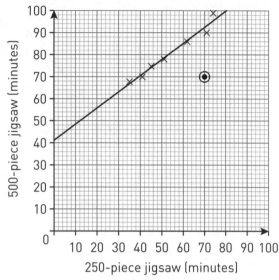

  b (70, 70). It does not fit the pattern of the other data.

  c Positive correlation. The longer the time taken to do the 250 piece jig saw the longer the time taken to do the 500 piece jig saw.

  d i 82 – 84 minutes

    ii Reliable for data as interpolation, but may not be reliable overall as small sample of students.

Designing questionnaires (page 64)

1 For example, ask students in the playground their gender and age. A questionnaire might ask:

What gender are you?	Boy ☐	Girl ☐

How old are you?

Under 8	☐
9 to 10 years old	☐
11 to 12 years	☐
13 or over	☐

Single event probability (page 65)

1 a $\frac{3}{10}$ b $\frac{7}{10}$ c 0

2 35% or 0.35 or $\frac{7}{20}$

3 $\frac{4}{7}$ 4 28

Combined events (page 66)

1 $\frac{17}{24}$ 2 $\frac{2}{19}$

Estimating probability (page 67)

1 a $\frac{14}{20}$ b 90 2 $\frac{29}{60}$

Mixed exam-style questions (page 68)

1 $\frac{3}{10}$

2 0.15

3 3064

4 a $\frac{10}{30} = \frac{1}{3}$ **b** $\frac{11}{30}$

$\frac{64}{105}$

0.032 to 3 d.p.

it is carried out outside the station, so it is biased towards people travelling by train.

b Question 1: overlap for aged 30. Improve by changing question to Under16, 16 to 29, 30 to 40, over 40.

Question 2: a vague question, doesn't give any idea of how many times, nor over what period of time, nor if single or return. Consider option boxes with number of return journeys in a month of 0 to 2, 3 to 5, 6 to 8, 9 or more.